Lifeways in the Northern Maya Lowlands

Native Peoples of the Americas

Laurie Weinstein, Series Editor

Lifeways in the Northern Maya Lowlands

New Approaches to Archaeology in the Yucatán Peninsula

EDITED BY
Jennifer P. Mathews and
Bethany A. Morrison

The University of Arizona Press

Tucson

The University of Arizona Press
© 2006 The Arizona Board of Regents
All rights reserved

11 10 09 08 07 06   6 5 4 3 2 1

Library of Congress Cataloging-in-Publication Data
Lifeways in the northern Maya lowlands : new approaches to
archaeology in the Yucatán Peninsula / edited by Jennifer P.
Mathews and Bethany A. Morrison.
        p. cm. — (Native peoples of the Americas)
     Includes bibliographical references and index.
     ISBN-13: 978-0-8165-2416-7 (hardcover : alk. paper)
     ISBN-10: 0-8165-2416-5 (hardcover : alk. paper)
     1. Mayas—Mexico—Yucatán (State)—Antiquities. 2. Mayas—
Agriculture—Mexico—Yucatán (State). 3. Mayas—Mexico—
Yucatán (State)—Politics and government. 4. Indigenous
peoples—Ecology—Mexico—Yucatán (State). 5. Social
archaeology—Mexico—Yucatán (State). 6. Yucatán (Mexico :
State)—Antiquities. 7. Yucatán (Mexico : State)—Environmental
conditions. I. Mathews, Jennifer P., 1969– II. Morrison,
Bethany A., 1969– III. Native peoples of the Americas
(Tucson, Ariz.)
F1435.1.Y89F53 2006
972′.6500497427—dc22
                                         2005028396

In loving memory of Enzo

Contents

Series Foreword, by Laurie Weinstein   ix

Introduction   3

Part 1   An Introduction to the Northern Maya Lowlands

1 Changing Archaeological Perspectives on the Northern Maya Lowlands   13
George J. Bey III

Part 2   Making a Living

2 From Swidden to Swamps: The Study of Ancient Maya Agriculture   41
Bethany A. Morrison

3 Cenotes, Wetlands, and Hinterland Settlement   56
Charles W. Houck Jr.

4 The Archaeology of Urban Houselots at Chunchucmil, Yucatán   77
Scott R. Hutson, Aline Magnoni, Daniel E. Mazeau, and Travis W. Stanton

Part 3   Ancient Politics and Interactions

5 Late Formative and Early Classic Interaction Spheres Reflected in the Megalithic Style   95
Jennifer P. Mathews and Rubén Maldonado Cárdenas

6 Foreign Lords and Early Classic Interaction at Chac II, Yucatán   119
Michael P. Smyth and David Ortegón Zapata

7 Classic Politics in the Northern Maya Lowlands   142
Justine Shaw and Dave Johnstone

8 Ichmul de Morley and Northern Maya Political Dynamics   155
J. Gregory Smith, William M. Ringle, and Tara M. Bond-Freeman

9 The Relationship between Tula and Chichén Itzá: Influences or Interactions?   173
Rafael Cobos

Part 4   Today's Scene

10   Ethnoarchaeology in the Northern Maya Lowlands:
     A Case Study at Naranjal, Quintana Roo   187
     Kurt R. Heidelberg and Dominique Rissolo

11   Archaeologists Working with the Contemporary
     Yucatec Maya   198
     Dominique Rissolo and Jennifer P. Mathews

12   Milpas of Corn and Tourism Milpas   210
     Alicia Re Cruz

     Bibliography   221
     About the Contributors   259
     About the Editors   263
     Index   265

Series Foreword

Native Peoples of the Americas is a multi-volume series that covers North, Middle, and South America. Formerly with Greenwood Press, and now with the University of Arizona Press, this series is unlike any other Native American series. Many series have a set format, and quite often, they are either annotated bibliographies about native cultures or they merely rehash and rewrite culture histories for introductory classes. Books in the Native Peoples series provide original research about the many culture areas that span the New World. Each volume broadly places the culture areas in time and space, with overviews of the archaeology and/or ethnohistory of the regions. Volume editors create the particular slant of the books, and each volume is unique. While some books examine the ethnogenesis of tribes, others describe gender relationships, resource use and competition, method and theory, ethnicity, environmental concerns, culture contact, and cultural survival. The books are written by a variety of scholars, including anthropologists, historians, and native peoples. In short, the volumes have something for everyone, from inveterate scholars to inquisitive college students.

This volume brings together a variety of approaches to Maya archaeology. While it emphasizes method, theory, and practice in Maya archaeology, *Lifeways* is unique because it examines the less-well-known northern Maya cultures, instead of the romanticized sites of the southern Maya region. The volume is divided into sections, consisting of chapters made up of related themes. The first part, "An Introduction to the Northern Maya Lowlands," introduces the reader to the environment and culture of the northern Maya lowlands (Bey). The second section, "Making a Living," includes chapters on the multi-disciplinary, multi-hypothesis approach to the examination of swidden agriculture (Morrison), hinterland/settlement pattern studies (Houck), and the examination of settlement in urban centers (Hutson et al.). The third section, "Ancient Politics and Interactions," presents a number of innovative chapters: Mathews and Maldonado Cárdenas present an argument for looking at the Megalithic architectural style to infer political interaction spheres; Smyth and Ortegón Zapata look at the possible influence of Teotihuacan on the site of Chac II; and Shaw and Johnstone propose the use of roadways, ceramics, and architecture to discern polities. Smith, Ringle, and Bond-Freeman apply territorial vs. hegemonic models of political control to the Maya data; and finally,

Cobos examines the ongoing debate about the relationship between the sites of Tula and Chichén Itzá. The fourth section, "Today's Scene," integrates archaeology with ethnography and includes chapters about the relationship of modern-day residential models to the archaeological record (Heidelberg and Rissolo) and about archaeologists as cultural agents and brokers in their research communities (Rissolo and Mathews). Finally, Re Cruz's chapter brings us full circle with the melding of the old and the new in her discussion of the tourism industry that is flourishing, thanks to archaeology on the Yucatán Peninsula.

This volume is also unique because it gives readers an appreciation of the science of archaeology and the many avenues that are utilized to ask, and then answer, questions about the past. As the volume editors suggest, "Science is not a recipe to be followed from beginning, to middle, to end. Rather, it is a process, a flowing of ideas, which sometimes takes as many steps backwards as forwards."

I think readers will agree that this book fills an important gap in the literature on Maya studies as it focuses on many areas not previously addressed. What began as a lunchtime conversation about research designs and hypothesis testing several years ago with Dr. Morrison has turned out to be an important contribution to northern Maya archaeology.

Laurie Weinstein

Lifeways in the Northern Maya Lowlands

Introduction

Jennifer P. Mathews and
Bethany A. Morrison

The collection of chapters in this volume originally stemmed from a symposium organized for an annual meeting of the Society for American Archaeology. During the transformation of that symposium into this volume, additional contributors were approached, including several of our colleagues from Mexico's Instituto Nacional de Antropología e Historia (National Institute of Archaeology and History), and several lofty goals were set.

The primary goal of the volume is to emphasize the heretofore-unrecognized importance of the northern lowlands (fig. I.1) in Maya prehistory. Second, we wanted to provide an overview of Maya culture —including aspects of subsistence, economics, social structure, and political history—appropriate for students, while incorporating current data that would be of interest to professionals. Third, we were encouraged by Laurie Weinstein, our series editor, as well as the current trends of our field, to bring a more human aspect to the work by including modern ethnographic data. And finally, it was our desire to illustrate the many faces of science by soliciting chapters from projects utilizing a variety of methodologies and theoretical approaches.

As George Bey notes in the introductory chapter, the research on the northern lowlands is notably absent from recent literature on the Maya area despite the fact that investigations have increased markedly in the last decade. All of the chapters in this volume reflect the wave of new information just now becoming available about the ancient inhabitants of the northern Maya lowlands. While much of the work in the area is preliminary, it is becoming abundantly clear that the northern lowlands were not merely the home to Postclassic refugees from the south, but rather the home of a culture that developed in place from the Formative period and that resulted in specific, complex adaptations to its social and physical environments. Nonetheless, these chapters should demonstrate the need for researchers in the north and south to ignore modern geographic boundaries and look to all of our colleagues for new ideas and innovations to further our understanding of the ancient Maya.

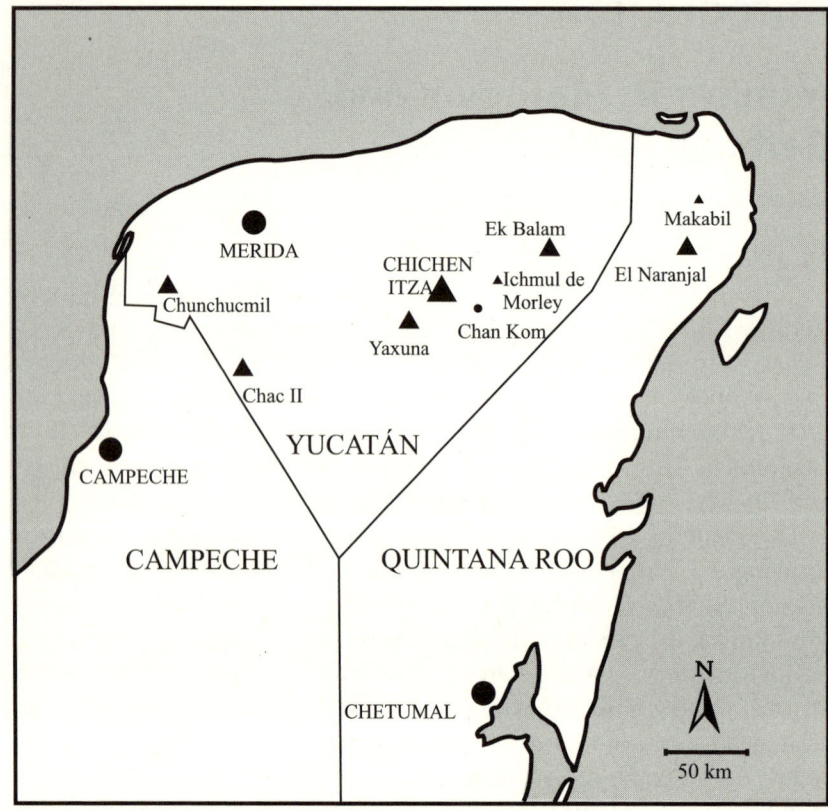

Fig. I.1 The major sites and locations discussed in the volume. Circles indicate modern settlements. Triangles indicate archaeological sites. (Map created by Bethany A. Morrison.)

The sections "Making a Living" and "Ancient Politics and Interactions" address our second goal of providing an overview of Maya culture along with current research. For example, Bethany Morrison's chapter on agriculture attempts to look at the question of how the Maya fed themselves within a wetland environment that differs greatly from those found in the southern Maya lowlands. A five-year study has revealed that the wetlands were manipulated to include extensive check dams that would have allowed the Maya to utilize the soggy environment to harvest resources such as palms, fish, or snails and even a natural fertilizer, known as periphyton. This algae contains freshwater mollusks and is high in nitrogen and phosphorus. The study tests whether or not this fertilizer may have been transported to home gardens in the nearby community of Makabil, by looking for the wetland mollusks in the homegarden soils, far outside of the wetland water line. Evidence of mollusks in the soils provides intriguing evidence that periphyton was used in this ancient community and has spawned further studies at other ancient sites located near wetlands.

Charles Houck, in his chapter on hinterland settlement, discusses the need to combine data from large sites with data from the surrounding areas to provide a more complete picture of the culture, including elite/non-elite and core/periphery studies. The intensive methods of hinterland survey allow archaeologists to incorporate environmental data with settlement data, allowing for a thorough understanding of how hinterland populations used their environment. Focusing on the hinterlands of Ek Balam versus the core area, Houck finds several trends that emerge from the data. One of the major findings is that populations found in outlying areas versus those found in the core may have been more complex than the anticipated elite/non-elite dichotomy. In fact, representatives of the ruling class may have lived in the hinterlands with areas of valuable agricultural lands, overseeing their development and ensuring the protection of important crops. This model emphasizes that this kind of detailed survey work takes the activities and contributions of all levels of society into account.

The chapter on Chunchucmil by Scott Hutson, Aline Magnoni, Daniel Mazeau, and Travis Stanton discusses the fact that, while this center lacks major monumental architecture and evidence for a centralized political structure, it contains the highest settlement density known for a Classic period site. This is particularly surprising, given that it is located in an area with marginal soils. This project has attempted to explain how such a dense population could have sustained itself, focusing in particular on the site settlement. They test the model first proposed by Hayden and Cannon (1982) that suggests the bounded settlements may represent corporate groups that resulted from economic and environmental stresses that forced multiple families to share resources. The authors conclude that while the bounded houselots appear to be corporate groups, they do not seem to form for the same reasons that have been suggested at other sites, such as defense and controlling trade networks.

The chapters by Jennifer Mathews and Rubén Maldonado Cárdenas; Michael Smyth and David Ortegón Zapata; Justine Shaw and Dave Johnstone; J. Gregory Smith, William Ringle, and Tara Bond-Freeman; and Rafael Cobos all look at the issues of political organization and "interaction" in the Yucatán Peninsula during the Late Preclassic and Early Classic, Classic, and late Terminal Classic periods. Each chapter highlights the methods that can be used to study different time periods to understand the dynamics of ancient sites. Mathews and Maldonado Cárdenas's chapter focuses on the interaction sphere model, looking primarily at an architectural style known as Megalithic. While this style is found at numerous sites, it has not been recognized as a widespread and contemporaneous style. Dating the architecture is challenging, as there are no associated texts like those commonly found at later sites, and the immense size of the architectural blocks has resulted in a poor ceramic sample. The authors discuss all known sites with Megalithic-style structures and their known dating, demonstrating a Late Preclassic and Early Classic interaction sphere across the north.

The chapter by Michael Smyth and David Ortegón Zapata argues for a possible influence from Teotihuacan at the site of Chac II in the Puuc Hills region, an area long thought to be out of the realm of this major central Mexican center. Using evidence from architecture, artifacts, mortuary patterns, and iconography, the authors argue that the Puuc region included major Early Classic settlements, and that foreigners were engaged in intense interaction in this region. They conclude that the strategic location of Chac II along overland trade routes, local resources, and the nearby water source in the Chac Cave would have been a major draw for foreign travelers, perhaps resulting in extensive trade, sharing of ideas, and even foreign residents at the site.

Justine Shaw and Dave Johnstone examine the physical traces of politics that were left behind during the Classic period. They note that while most sites in the northern Maya lowlands lack the abundant hieroglyphic texts found in the south, there are plentiful ceramics that can not only help define dates but can reveal information about the politics of the users. They emphasize that rather than having broad ceramic spheres, northern ceramic complexes appear to be more local and tied in with historical events. Like Mathews and Maldonado Cárdenas, they examine shared architectural styles across sites as an indication of shared influence. Finally, they look at the roads or *sacbeob* that represent the literal connections between sites as a way of understanding political control. Close examination of these remnants reveals that the trajectory of politics in the north was not so different than the south, and that sites like Yaxuná had divine rulers who appear to have participated with southern cities.

J. Gregory Smith, William Ringle, and Tara Bond-Freeman stress the biases of research related to Maya politics, such as the failure to include northern sites in the general discussion of Maya political organization and an over-emphasis of single-site analysis versus a regional approach. Focusing on the regional approach, the authors argue that using the unitary-state and segmentary-state models of political organization as mutually exclusive, rather than poles of the continuum, is a limited way of examining Maya politics. Looking at the site of Ichmul de Morley—a border site located between the major centers of Ek Balam and Chichén Itzá—they test what kind of influences from each site were found at the smaller center to determine the type of political control that was wielded. They conclude that despite the pivotal location of Ichmul de Morley, neither of the major centers exhibited an obvious presence. The authors argue that the results of this study may indicate a weakness in how we conceptualize borders and the transfer of political influence, ideas, and goods from the core to the periphery.

The chapter by Rafael Cobos looks at the age-long debate of the relationship between the Toltec of Tula and the Maya of Chichén Itzá. Examining the "migration, invasion, and conquest" model that has dominated academic arguments during much of the twentieth century, he breaks down the evidence for a Toltec invasion and conquest of the Chichén Maya. Refuting evidence such as historical documents and sculptural, ceramic, and lithic remains that argue for this one-way domination, Cobos argues that instead the "local development" model may better fit the data. He concludes that this conflation of the two styles may be explained better through interaction and local development than through domination and one-way influence.

The section entitled "Today's Scene" addresses our third goal, to highlight the important "human" dimension of archaeology in the northern lowlands—the regular contact between native peoples and the researchers that work there and the benefits and difficulties that can stem from this interaction, as well as the incredible impact of tourism in the area. While perhaps not traditionally emphasized in archaeological volumes, Maya archaeology is beginning to embrace and recognize the value of examining these issues.

For example, Kurt Heidelberg and Dominique Rissolo's chapter provides a useful model for studying the modern Maya to better understand ancient Maya houselots in the archaeological record. This complex environment combines the areas of home, garden, workspace, and orchard, making it difficult to define boundaries of these activity areas. Using Tom Killion's model of the Household Garden–Residence Association, Heidelberg and Rissolo base an ethnographic study on houselots in the modern community of Naranjal to understand how the contemporary Maya manage space. While the authors note that caution must be used in making comparisons between the ancient and modern Maya, they find that we should be able to anticipate some general patterning. This innovative model allows researchers to reconstruct the activity areas, even if the physical features are lacking.

In the chapter by Dominique Rissolo and Jennifer Mathews, the authors examine the issues of archaeologists living and working in modern Maya communities, a situation that differs greatly from non-community projects. Based on personal experiences, they examine the responsibilities that archaeologists must accept by choosing to conduct field projects within local communities, including the possibility that community members may impact research designs, control access to sites and artifacts, or request contribution to, and participation in, community events. The authors also consider that it is ultimately the people of the village who are the agents of change and who control their own cultural patrimony.

Alicia Re Cruz's chapter examines the impact of tourism on the modern Maya, focusing in particular on the town of Chan Kom. Due to its proximity to the major archaeological site of Chichén Itzá, this village has been subjected to archaeological and ethnographic researchers during much of the twentieth century. Best known from Redfield and Villa Roja's ethnographies, ReCruz explores the more recent evidence that demonstrates that this important pueblo has been socially fragmented by the development of tourism. Out-migration to Cancún has resulted in some community members garnering some wealth and political power. They also recognize the value of "authentic Maya culture" to tourists and hope to capitalize on this by bringing tourism to the village. Those who have not left Chan Kom are resistant to these changes, as they feel these migrants have lost their ties to the *milpa* and what it means to be Maya.

Finally, overall, this volume demonstrates the many shapes of science in archaeology. The chapters are formatted to highlight research questions, hypotheses, and interpretations—the fundamental steps in the scientific method—but the reader will note that each research project approaches these steps differently. Science is not a recipe to be followed from beginning, to middle, to end. Rather, it is a process, a flowing of ideas, which sometimes takes as many steps backwards as forwards. As any good scientist knows, but often forgets to convey to his or her students, science is largely inductive. The true brilliance of science is in the creation of hypotheses or ideas to be tested. Some research presented here is still in its preliminary stages, other projects are more established, but each, in the process of analysis, has discovered new questions to be asked. Through example, this volume highlights not just the methods but also the process of scientific inquiry. It is this lesson that we feel may be our most significant offering to those who read this book.

Acknowledgments

The editors would like to thank several people for making this volume possible: first and foremost, the modern Maya of Yucatán with whom many contributors in this volume have worked. We thank you for your friendship and for sharing your cultural past. Without their support of our research, none of this would be possible. We would also like to thank the Instituto Nacional de Antropología e Historia for their continued support of research in the region, and, in particular, the archaeologists working in the offices in Mérida, Cancún, and Chetumal. Additionally, we want to thank Laurie Weinstein, the series editor, for initiating this volume; Allyson Carter and Christine Szuter of the University of Arizona Press for their support of this project; Ashley Fry and Andrea Rodriguez for their editorial assistance; and the anonymous reviewers for making this a better volume. We would especially like to thank Traci Ardren for her thoughtful comments on earlier versions of this manuscript, as she greatly shaped and improved the volume. Finally, we would like to thank our families and loved ones for their undying support through long nights at the computer and even longer seasons in the field; it means everything to us to know they're behind us.

Part 1 An Introduction to the Northern
Maya Lowlands

1

Changing Archaeological Perspectives on the Northern Maya Lowlands

George J. Bey III

Across the vast expanse of the northern Yucatán Peninsula, one of the great expressions of pre-Columbian civilization developed, for it was here that a major regional development of the ancient Maya took shape (fig. 1.1). Although strongly connected to the larger Maya world and to the even greater universe of Mesoamerican culture, it was also a world unto itself with a distinct history and culture. Despite over 150 years of research, it is only recently that this point has become clear to Maya archaeologists. As a result, the northern Maya lowlands remained one of the least understood and most marginalized areas of the Maya world. Due to the history of Maya archaeology and to the nature of the archaeological remains, our understanding of the northern Maya lowlands was biased in terms of its overall place in Maya prehistory.

Fortunately, the past twenty-five years have produced an explosion of archaeological research in the northern Maya lowlands, resulting in a major revision of the prehistory of this region. Large-scale projects were undertaken at sites across the peninsula, including Aké (Maldonado 1980, 1989), Chichén Itzá (Cobos 2004; Schmidt 1999, 2000), Chac II (Smyth 1998, this vol.), Chunchucmil (Dahlin and Ardren 2002; Hutson et al., this vol.; Stanton et al. 2000), Cobá (Benavides Castillo and Manzanilla 1987; Con and Martinez Muriel 2001; Folan et al. 1983; Robles Castellanos 1990), Dzibilchaltún (Maldonado et al. 2001; Repetto Tío 1986), Ek Balam (Ringle et al. 2004; Vargas de la Peña and Castillo Borges 1999, 2001; Vargas de la Peña et al. 1999), Komchen (E. Andrews V 2003; Ringle 1999), Izamal (Maldonado 1990; Millet and Burgos Villanueva 1998), Labná (Gallareta Negrón et al. 1999), Mayapán (Milbrath and Peraza Lope 2003; Peraza Lope 1999), Oxkintok (Rivera Dorado 1991), Sayil (Sabloff and Tourtellot 1991; Tourtellot and Sabloff 1994), Uxmal (Barrera Rubio and Huchim Herrera 1989; Kowalski et al. 1996), Xcambó (Sierra Sosa 1999, 2001), Xculoc (Miche-

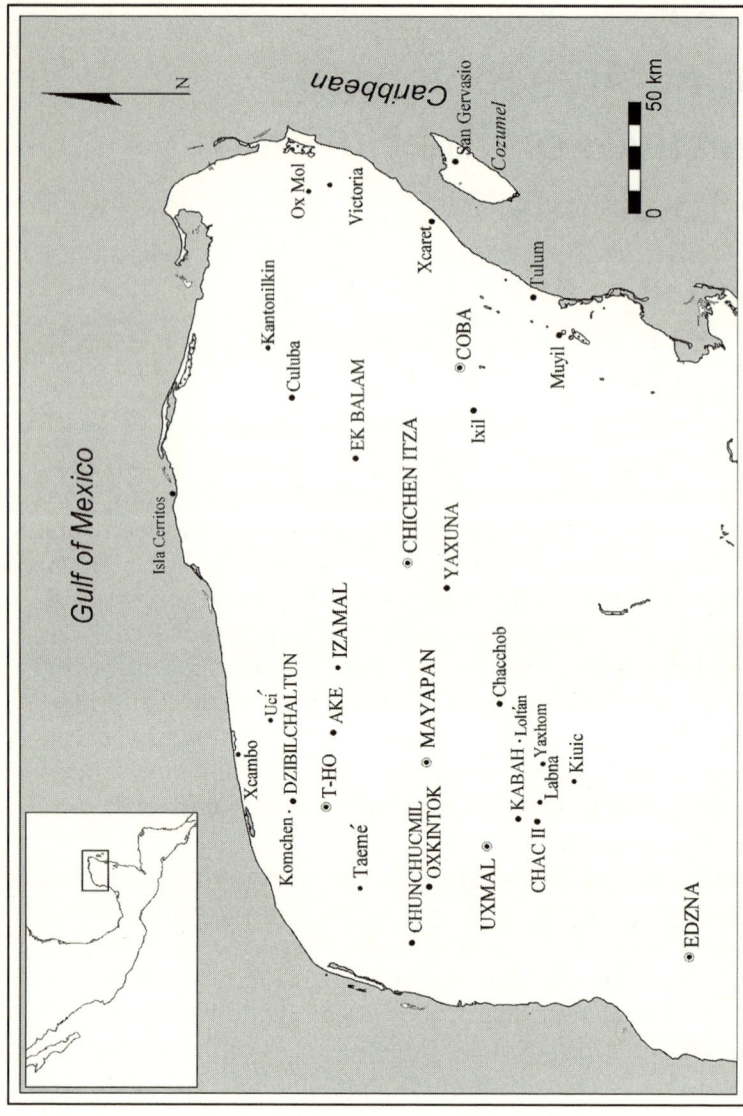

Fig. 1.1 Northern Maya lowlands with sites mentioned in text. Sites in capital letters represent major centers in the north, while those with circles around the dots indicate the best known of these sites. Secondary and tertiary sites are indicated by names shown without all caps. (Figure created by George Bey III.)

let et al. 2000), Xkipche (Vallo 2000), and Yaxuná (Ardren 1997; Freidel 1986; Shaw and Johnstone 2001; Stanton 2000; Suhler 1996). In addition, the north has recently seen a number of intensive regional surveys: the Chikinchel Project (Kepecs 1998), Cupul Survey (A. Andrews, Gallareta Negrón, et al. 1989), Ek Balam Project (Ringle et al. 2004; see also Houck, this vol., Smith et al., this vol.), Puuc Hills Project (Dunning 1992), Yalahau Project (Fedick and Taube 1995; J. Mathews 1998; Morrison 2000; Rissolo 2003; see also Heidelberg and Rissolo, this vol., Mathews and Maldonado Cárdenas, this vol.) and the Costa Maya Project (A. Andrews and Robles Castellanos 2001), as well as numerous salvage projects, and smaller exploratory projects. Much of this work has been carried out by Mexico's Instituto Nacional de Antropología e Historia (National Institute of History and Anthropology); however, a significant amount of the research has been multinational with projects from Spain, France, Germany, and the United States. National and international teams have investigated some sites, such as Ek Balam, simultaneously.

A result of this research has been a dramatic reevaluation of many of the basic positions in the thinking about this region. The cultural history of the north has been revised, with concomitant changes in the traditional views on chronology, political history, economics, art, and iconography, as well as regional and extra-regional interaction and influence. Some of the specific points that have been the focus of this rethinking are: (1) the nature of the Formative (Preclassic) occupation and the transformation of Formative to Classic culture; (2) the size and scale of Early Classic Maya civilization in the north; (3) the similarities and differences between Classic Maya institutions and cultural forces, such as warfare and trade, in the north and the south; (4) the nature of the Terminal Classic transformation in the north, especially as regards Chichén Itzá; and (5) the continuity and change between Classic and Postclassic society. While space will not permit me to examine all of these points, this chapter will focus on possible reasons why the north has been underrepresented in the literature, as well as examine some of the new thinking related to the northern Maya lowlands, in particular during the Formative and Early Classic periods.

Questions/Problems: The Northern Lowlands in Maya Prehistory

The traditional view of the northern Maya lowlands can be summarized fairly easily. It is held that although there was some occupation in the north beginning in the Late Formative (~400/300 BC–AD 250/

300), cultural development was limited until the Late and Terminal Classic periods (AD 700–1050), when the north exploded. The view was that as the southern Maya world was beginning its decline, population movement and cultural florescence headed into the northern Maya lowlands with the occupation of the Puuc region, Dzibilchaltún, and Cobá. This was followed by the emergence of Chichén Itzá in the Early Postclassic period (AD 1050–1200), the rise and fall of Mayapán, and the development of the decadent Late Postclassic world (AD 1200–1500s), typified by Tulum. What this story was missing was the development of a great in situ civilization with polychromes, literacy, and Classic culture. By default, the Terminal Classic period was seen as the apogee of the north. Archaeologists familiar with the north recognized that northern prehistory was more complex and significant than was generally held (Ball 1977); however, their views had little impact on the overall place of the north in Maya studies.

That the research of the northern Maya lowlands failed to make an impact on many of the major ideas we have about the ancient Maya becomes obvious when we look in most textbooks and edited volumes related to the Maya. For example, one of the most famous and commonly used textbooks of our field, *The Ancient Maya* (Sharer 1994), mentions the northern Maya lowlands only briefly for the Formative period and not again until the Terminal Classic period. It is noted that there are several Early Classic sites, but they are poorly known and have produced few texts, and no major centers existed until the Terminal Classic period. Even the most recent textbooks take virtually the same perspective (see Demarest 2004). A perusal of major edited volumes on Classic Period Maya art, architecture, and politics also reveals the same lacunae. Unfortunately, many authors think that it is not until the Terminal Classic period that the north is worth examining.

The history that produced this view is too complex to fully explore here; however, there are a number of general factors that can be seen as playing significant roles in establishing this perspective. One of the most important facts contributing to the idea that the north did not play as central a role as the south is the simple fact that "cultural sequences of this region until recently have included very little information about the periods before the Late Classic and the Puuc phenomenon" (Varela Torrecilla and Braswell 2003). This was a result of the limited amount of archaeological research in the north that focused on the earlier time periods. Obviously, this research bias in turn contributed to several of the misconceptions that plagued our understanding of the region.

One of these misconceptions was that the north lacked dense occupation or highly developed centers in the Early Classic period. Based on the work of E. Andrews IV, E. Andrews V, and their colleagues at Dzibilchaltún and Komchen (see, for example, E. Andrews V 1981, 1988; E. Andrews IV and Andrews V 1980), other archaeologists have generalized the large-scale collapse at the end of the Late Formative period at these sites to the rest of the north and comfortably considered that the north did not recover until the rise of the Late Classic and Terminal Classic states, particularly with the explosion of new sites in the Puuc. Ball stated that "the collapse at Dzibilchaltún and Komchen did not appear to have been symptomatic of a general Northern Plains condition" (1977:120) and several important sites were long recognized as having significant Early Classic occupations (Aké, Acanceh, and especially Izamal), but as they were poorly known (Mathews and Maldonado Cárdenas, this vol.), they were largely unconsidered. The glory of the northern Maya was thus late, and in the areas of most powerful expression, like the Puuc, was thought to be largely without in situ development; and, though architecturally impressive, essentially lacking in the sophisticated art and literacy of the south. Thus, the time period gets classified as *Terminal* Classic; a term that I argue is an artifact of the southernocentrism that dominates Maya archaeology.

Another contributing factor was that the north also suffered from lack of coverage. For example, the area that includes the Chikinchel region, the Ek Balam region, and the Yalahau region, as well as a large number of important centers, was virtually unknown as late as the 1980s and 1990s. Interpretations about these blank spots on the map were either not discussed or were sometimes generalized as being essentially the same as the better-documented areas.

A third factor is the apparent lack of historic documentation during the Classic period. Whereas since the 1970s the southern lowlands have emerged as a world of dynasties and kings, of events and stories, the north for the most part has remained mute. The questions we ask ourselves in the north are: a) to what degree is this epigraphic silence archaeological; and b) to what degree is this epigraphic silence cultural?

As Shaw and Johnstone point out (this vol.), part of this problem in the north appears to be the result of the nature of our archaeological remains. The limestone on which texts were carved is soft and easily erodes, which means that many texts, if they ever did exist, are gone. Also, I would note that plain stelae are not uncommon on sites in the northern Maya lowlands, suggesting that texts in some cases were painted on stucco surfaces that have long since disappeared. Although this may explain part of the lack of history, we are still left with the fact that overall there are far fewer stelae and inscriptions, eroded or not, in the north than in the south. Of those we do have, the vast majority date to the Late and Terminal Classic periods and are short texts. A total of 80 percent of the known texts in the north deal with dedication and are found on lintels and capstones. In dating, style, and linguistics, they represent a different scribal tradition than that associated with the southern lowlands (Grube 2004). The dearth and type of texts have significantly contributed to the idea that the north was a cultural backwater during the glory days of the Maya Classic period.

Additionally, there is the lack of an elaborate Classic polychrome ceramic tradition such as is found in the south. Although there are some types of polychromes produced in the north, the Early Classic ceramic tradition consists primarily of bichromes and monochromes, with little evidence of texts and a general lack of artistic sophistication. The Early Classic ceramic tradition evolves into the slate-ware tradition that forms the bulk of the Cehpech ceramic sphere, where monochrome continues to dominate the ceramic wares, with elite pieces instead defined by thinness, incision, and carving. The great painterly tradition of polychromes never appears in the north.

A final reason for the north's minimal role in our thinking of Maya culture prior to the Terminal Classic period is the relatively low level of publication by researchers working in this area. We must publish our results more fully and in a more timely fashion than we have in the past. Huge amounts of information from many major projects, some decades old, remain unpublished. I am not sure if other areas are as guilty of this as we are, but I know it is a major issue in the archaeology of the northern Maya lowlands. Taking these facts together, it is not surprising that the northern Maya lowlands were perceived as they were. Fortunately, with an increased understanding of the nature of cultural evolution in the north, the scale of occupation, and the recognition that it was an in situ development with Formative and Early Classic roots and history, this perception is changing.

Changing Views of the Formative and Early Classic Periods in the North

The Middle Formative Period

There has been a dramatic transformation in our understanding of the Formative world in the northern Maya lowlands since the Komchen Project in 1980 (E. Andrews V 2003; Ringle 1999). Although the existence of Middle (~800/700–400/300 BC) and Late Formative period (~400/300 BC–AD 250/300) occupation in the north was reported well before E. Andrews V's work at Komchen (see Ball 1977 for a summary;

Brainerd 1958), it was his efforts that established the nature and scale of Middle and Late Formative Maya culture in the northern Maya lowlands. E. Andrews V and his colleagues defined the evolution of settled village life beginning perhaps as far back as 700 BC (E. Andrews V 1988, 1989, 1990, 2003; Ringle and Andrews 1988). Their settlement work and their excavations provided us with an understanding of the size and growth of a Formative community and a view of both domestic and political/ceremonial architecture and culture. In addition, E. Andrews V's analysis of the ceramics of Komchen resulted in the first systematic, detailed definition of a sequence of Formative ceramic complexes in the northern Maya lowlands (E. Andrews V 1988, 1989). His research not only defined the types and varieties of pottery associated with the various time periods represented by these complexes, but also highlighted important typological issues in the understanding of Formative pottery in the north.

In the past two decades, our understanding of the Formative occupation of the north has significantly increased. Where Komchen was once one of the few well-known Late Formative sites in the north, it is now recognized that Late Formative culture was spread across the entire northern peninsula. And, where once, Komchen was just about the only true Middle Formative site to be identified (though Middle Formative ceramics were known from elsewhere in the north [Boucher 1991]), there is now evidence that the Maya occupation of the north during this time period was also widespread.

Evidence of Middle Formative (~800/700–400/300 BC) communities is found at Komchen (E. Andrews V 1988; Ringle and Andrews 1988), Ek Balam and the small nearby site of Xuilub (Bey et al. 1998; Ceballos Gallareta 2004), and Yaxuná (Ardren and Johnstone 1996). In the northwest corner of the peninsula, the Costa Maya Project (A. Andrews and Robles Castellanos 2001) has identified 116 Middle Formative sites, many of them with ballcourts, including the proposed regional center of Xtobó (D. Anderson 2004; Robles Castellanos in press). In the Puuc region, significant Middle Formative occupations have been defined for Kiuic and for a newly identified site near Labná called Paso del Macho, which has a ballcourt (Bey and May C. 2005; Gallareta Negrón et al. 2003). The presence of Middle Formative occupation is also known from Labná, Loltún Cave, and Maní Cenote, again in and near the Puuc (Boucher 1991). Other major Middle Formative settlements known from the region between the Puuc region and Mérida include Xocnaceh and Poxila (Gallareta Negrón and Ringle 2004; Robles Castellanos in press). Middle Formative material has also been recovered from Tipikal (Peraza Lope et al. 2002), Caucel, Aké, Izamal, Mayapán, Acanceh, Isla Cerritos, and Cobá (Hernandez n.d.:42). Recently, surface collections and excavations in caves found in the Yalahau region of northern Quintana Roo have "produced a moderately sized but nearly complete assemblage of Middle Formative ceramic groups" (Rissolo et al. 2005; Rissolo and Ochoa Rodriguez 2002).

In the case of many of these sites, the Middle Formative occupation is associated with monumental architecture. At Kiuic a sequence of construction in the Yaxché group includes a 1 m high platform that has a minimum dimension of 28 × 28 m. The remains of a 14 m long structure have been defined on the east side of this platform along with a second more elaborate structure on the south side (Gallareta Negrón et al. 2004). This structure, N1015E1015-sub, is at least 13 m long, with a stucco covered *talud* and rounded corners. At Yaxuná, significant amounts of Middle Formative ceramics were recovered from Structure 5E-19, a triadic cluster of mounds (Ardren and Johnstone 1996; Suhler et al. 1998). They note that the highest, 5E-19, rises 6 m above the 6 m high supporting platform and that three Late Formative construction phases containing Middle Formative materials were encountered in the upper 2 m of the deposit, indicating a strong possibility of substantial Middle Formative supporting structures.

Equally provocative are the many Middle Formative sites located by A. Andrews and Robles Castellanos (2003) as part of their Costa Maya Project. Surveying the northwest coast of the peninsula, they have identified twenty plus Middle Formative sites that include formal monumental architecture and ballcourts. The largest of these sites, Xtobó, also includes a number of *sacbeob* (roads) connecting groups of monumental architecture to a central plaza group (D. Anderson 2004; Robles Castellanos in press). The site of Paso del Macho, located in the Bolonchen District of the Puuc region (Gallareta Negrón and Ringle 2004), is another Formative site with a ballcourt and formal monumental architecture. Test pits indicate pure Middle Formative deposits in stratigraphic context (Chris Gunn, personal communication 2003). The most impressive monumental architecture is from the site of Xocnaceh (Gallareta Negrón and Ringle 2004), located along the northern edge of the Puuc escarpment, and from Poxila, 65 km further northwest of Xocnaceh (Robles Castellanos in press). At Xocnaceh, Gallareta Negrón has identified a spectacular Middle Formative acropolis. This 150 m × 150 m platform stands 8.5 meters above the surface and reached its maximum size during the final part of the Middle Formative period. Except for Late Formative floors laid down over the final Middle Formative construction phase, the entire basal platform appears to date from this early time period (Gallareta Negrón and Ringle

2004). The platform supports a number of other buildings that probably are Middle Formative in date. This is the largest Middle Formative
structure known in the northern Maya lowlands. The main construction at Poxila is equally impressive. It consists of a 2.5 m high basal
platform 100 m east-west × 90 m north-south. This platform forms an
acropolis with a huge structure on its east side. This structure is an 80 m
north-south × 40 m east-west platform that rises an additional 10 m
above the surface. Excavation of the structure has revealed that its various construction phases date almost exclusively to the Middle Formative period (Robles Castellanos in press). The construction techniques
used in both the Poxila and Xocnaceh platforms are basically identical, characterized by the use of large stone blocks.

At present, no Early Formative occupation is known from the northern Maya lowlands, and so archaeologists ask how and why did this
widespread, complex Middle Formative occupation of the north take
place. The two major theoretical models that address these questions
are those of E. Andrews V (1990, 2003) and Stanton (2000). They both
depend on differing interpretations of the Middle Formative ceramics in the northern Maya lowlands as reflections of sociocultural processes. E. Andrews V has interpreted the Middle Formative data as reflecting the arrival of the first sedentary villages in the northern plains.
His interpretation is based on the results of a detailed comparative
analysis of contemporary ceramic complexes from the Petén and Chiapas and the fact that ceramics found at late Middle Formative sites
identified by the Costa Maya Project in the northwestern corner of the
Yucatán are the same as those from Komchen (E. Andrews V 2003:5–6).
The donor area includes the sites of Altar de Sacrificios, Seibal, Nakbé,
and El Mirador. E. Andrews V suggests "this entire northwest corner of
the Yucatán Peninsula was settled in a short span of time by sedentary
farmers whose pottery links them directly to the southwest Petén and
Chiapas" (E. Andrews V 2003:6).

We now know that Middle Formative ceramic complexes are found
across the northern Maya lowlands, and that they exhibit some regional variation. E. Andrews V (2003:7) sees this as likely reflecting the
fact that "the destination of stimuli and probably immigrant groups"
into these areas came from the eastern Petén and Belize. He suggests
again then that the Middle Formative occupation in other parts of the
northern Maya lowlands resulted from ceramic-using migrants who
relocated in the area from further south. In an earlier article, E. Andrews V (1990) also argued for a later eastern migration originating in
northern Belize and the northeast Petén, which brought much of the
Classic Maya ritual complex up into Quintana Roo.

He did so because at that time he also believed that the area lacked a significant Middle Formative occupation (E. Andrews V 1990:15). Given that he now acknowledges the widespread distribution of Middle Formative ceramics across the northern Maya lowlands, E. Andrews V may no longer see a need for a later migration.

Stanton points out that underlying this model is the idea that the northern Maya were not culture innovators, but just waves of immigrants with a culture that originated in the south and spread north. He is unconvinced that the present lack of evidence for a pre-Mamon (pre–Middle Formative) occupation necessarily indicates that there was none. He also asks that if there were pre-Mamon people in the north, could not "these initial Yucatec populations have begun to make the transition to a sedentary way of life by emulating their southern neighbors?" (Stanton 2000:9). Could the Middle Formative complexes represent, instead of migration, the adoption of ceramics by an already existing population? Part of his argument rests on the understanding that the Early Nabanche and northern Mamon complexes are regionally distinct from those to the south, and "if populations migrated into the northern Maya lowlands from the south, why did they not bring their exact ceramic tradition?" (Stanton 2000:11). This is not necessarily a strong argument against migration since it would mean that migration equals cultural stasis, and in fact, some change would be expected as pottery producers adapted to local needs and materials. Ultimately, he does not deny the migration model of E. Andrews V may be correct. However, he insists that there is room for an alternative model—in this case his emulation model—although, he concludes by stating that "it is likely that the situation is more complex than either of these two explanations lay forth" (Stanton 2000:11).

The question of the existence of pre–Middle Formative occupation in the northern Maya lowlands is an important one. A lack of evidence led Ball (1977) to assert that as late as 700 BC the northern lowlands were uninhabited; however, today many northern Maya archaeologists believe a pre-Middle Formative occupation will be found and that "there were groups of hunter-gatherers and probably farmers in northwest Yucatán before 700 BC" (E. Andrews V 2003). Anthony Andrews (personal communication 2003) thinks it likely that the northern peninsula was occupied since Paleo-Indian times. He strongly suspects that Archaic and Paleo-Indian remains will eventually be found on the old beach ridges behind the coast and elsewhere. There is a date of 8250+ BP from Carwash Cave near Tulum (Coke et al. 1991), associated with a possible underwater cave hearth. The caves in the Puuc are another likely locale, he notes, pointing to Loltún Cave material that seems to have a Late Archaic component. Unfortunately, the evidence for Paleo-Indian occupation there is not based on any real solid context or dating (despite Velázquez Valádez 1980).

But, if such early populations existed, what was the relationship between the Maya using Middle Formative pottery and early monumental architecture and these original inhabitants? Is the change a result of diffusion, or migration, or some more complex model, as Stanton suggests? If E. Andrews V (1990, 2003) is correct, we are looking at a complex process in which two or more distinct Maya populations are coming into contact in the north. Is it likely that the pre-pottery populations absorbed the migrating social groups or were absorbed by them? Was there displacement of some sort by the migrants in some early clash of Maya cultures? If Stanton is right, we would be looking at the transformation of an indigenous population who were reacting to stimuli in the south but were modifying them to meet local needs and cultural norms. The search for a pre-Nabanche (pre–Middle Classic) occupation is an important research goal for archaeologists in the north, for only by finding evidence of one can we hope to fully evaluate the positions of E. Andrews V and Stanton. For many years, archaeologists thought that Middle Formative ceramics were restricted to the northwest portion of the Yucatán Peninsula. Now, we recognize Middle Formative occupations everywhere in the north. I believe we will soon begin to recover material from the pre-Nabanche period in the north, and I believe it will include an Early Formative ceramic component.

Even if the Middle Formative complexes do represent a migration into the northern lowlands, all indications are that Late Formative complexes evolve directly out of them. The Ek ceramic complex may be intrusive at the point of the Middle to Late Formative transition at Komchen (E. Andrews V 1988, 1989); however, it has not been defined elsewhere yet, and if you remove it from Komchen, the Late Nabanche clearly evolves from the Early Nabanche. At the present time, there seems little that links the Ek complex with the general evolution of ceramics in the north. The larger picture suggests an in situ evolution of northern ceramics complexes onward from the Middle Formative.

Stanton (2000) also argues for the emergence of social stratification during the Middle Formative period in the north, particularly at Yaxuná, on the basis of the identification of non-locally-produced pottery in the form of unspecified white-slipped ware and orange-slipped ware. The characteristics of the trade wares are not well established, and other than their non-local status, there is little exact information presented about them. What is significant, though perhaps not as sexy, is the simple fact that trade wares are found in the Middle Formative period. Besides the Yaxuná pottery, Muxanal group ceramics found at Komchen (E. Andrews V 1988, 1989), Ek Balam (Bey et al. 1998), Xocnaceh (Gallareta Negrón, personal communication 2005), and sites in the Puuc such as Kiuic, Labná, and Paso del Macho are also considered examples of long-distance trade ware. Given the limited evidence for long-distance exchange of ceramics during the Middle Formative period in Mesoamerica, this is exciting news. One also needs to consider what qualifies as evidence of long-distance trade during the Middle Formative period. Seemingly common locally produced ceramic types might have been exchanged occasionally over distances of 60–100 km and may represent the majority of ceramics moving across the regional landscape. However, without detailed petrographic and preferably compositional analysis, it is likely that these "trade items," if they exist, will not be identified. The study of the distribution of lithics and shell should also provide clues as to the nature of regional exchange systems in the northern Maya lowlands. Evidence from Xocnaceh indicates that shell was finding its way inland in significant quantities during the Middle Formative, with almost all of it coming from the north and west coasts of the peninsula (Cobos 2005). There was little evidence for shell from the eastern side of the peninsula and no evidence for material from further afield (Cobos 2005).

Contributing to this discussion are other kinds of evidence, both from the northern Maya lowlands and elsewhere, for long-distance trade. Although there is no evidence of pottery from the Gulf Coast being traded into these northern Maya sites during the second half of the Middle Formative period, Early Nabanche pottery has been recovered from Olmec sites dating to this time period. Early Nabanche ceramics, in significant amounts for trade ware (about 2% of the total collection), have been recovered from well-dated stratigraphic excavations at San Andrés, Tabasco. The deposits containing Early Nabanche ceramics date to the second half of the Middle Formative period, making them contemporaneous with the events going on during this time in the northern Maya lowlands (Von Nagy et al. 2002). This corroborates earlier identification by E. Andrews V (1986) of Early Nabanche pottery in pottery collections from La Venta, Tabasco, and Tres Zapotes in Veracruz.

Despite a lack of Olmec pottery from the Tabasco-Veracruz region, it is becoming increasingly clear that goods from that area were finding their way into northern Maya lowlands sites during the Middle Formative period. In addition to the famous Chacsinkin jades recovered from a Late Classic deposit, but considered to be Olmec in origin

(E. Andrews V 1986), recent work has discovered jade and green stone artifacts both at Poxila (Robles Castellanos in press) and Tipikal (Peraza et al. 2002). Beside jade, basalt was also being traded into the northern Maya lowlands at this time. Pieces of worked basalt are now known from Middle Formative deposits at Xocnaceh, Paso del Macho (Gallareta Negrón and Ringle 2004), and Kiuic.

The evidence for substantial public and ceremonial architecture, site hierarchies, regional exchange, and contact with the Gulf Coast region during the Middle Formative provides strong support for arguments of considerable social complexity during this time period. Stanton (2000) has argued that it was during this time that we begin to see elite alliance formation. Robles Castellanos (in press) argues for even greater complexity, suggesting that we are witnessing the emergence of archaic states in the northern Maya lowlands. However, Ringle (2005) suggests that we should be cautious in applying terms like "state" and "chiefdom" to the archaeological record. Although it is possible archaic states emerged in the Middle Formative, the presence of large platforms and ball courts do not necessarily prove their existence. Instead, he offers the idea that what we might be seeing is the development of mechanisms by egalitarian societies to manage the social issues that were arising in the Middle Formative with the increase of population both locally and regionally. In his view, they are likely the precursors to ranked or stratified societies rather than, as Robles Castellanos and Stanton suggest, the evidence for them.

Regardless of the specific level of social complexity, this new evidence makes it harder for archaeologists like Stanton to believe that migration can fully explain the Middle Formative cultures of the northern Maya lowlands. The number of sites across the northern peninsula, the impressive size of such sites as Xtobó, Xocnaceh and Poxila, the scale of ceremonial architecture, and the building of ballcourts all indicate that a widespread, highly organized and potentially complex society, fully in line with what was to become "the Classic Maya ritual complex," was already in place during the Middle Formative.

The Late Formative

The Late Formative (~400–300 BC–AD 250–300) in the northern Maya lowlands is marked by growth, change, and increasing regionalism. It is now clear that there is evidence of substantial Late Formative occupation throughout the northern Maya lowlands. All of the areas and most of the sites discussed above, including those in the Puuc, continue to be occupied in the Late Formative, although in a significant number of cases, such as Poxila and Xocnaceh, the Late Formative occupation consists of little more than renovation of already existing Middle Formative structures. Regional surveys also indicate that many Classic period secondary centers have at least Late Formative roots and that rural occupation was extensive during this period. Such surveys of the Chikinchel and Ek Balam regions show that well over half the sites identified in each region were occupied during the Late Formative (Bey et al. 1998; Kepecs 1998), as were the majority of sites surveyed in the Yalahau region (Andersen 2001; Fedick and Taube 1995; J. Mathews 1998; Morrison 2000).

One of the changing perspectives on Late Formative settlement is the increasing evidence of population in the Puuc. A summary of research by Dunning in 1992 concluded that although "a significant population appears to have occupied lands fringing the Puuc . . . the Puuc would seem to have been relatively lightly populated on a permanent basis during the Late Formative and the nature of that occupation remains problematic" (Dunning 1992:64). With the addition of more recent research, such as that of the Labná-Kiuic Regional Archaeological Project, this view is changing. The latest findings suggest there was widespread and substantial Late Formative settlement in the Bolonchen region (Gallareta Negrón et al. 2002, 2003). It is also becoming clear that the lack of Late Formative occupation in the Puuc was partially due to the history of research in this region. Archaeological work focused on monumental architecture in the site cores, where it is unlikely that one will encounter Late Formative occupation without extensive excavation beneath monuments and into platforms and plazas. Although we must withhold final judgment at the present time on the scale of Late Formative occupation in the Puuc, it is likely that many Puuc centers have substantial Late Formative occupations buried beneath later construction or in areas outside the site centers. It is also likely that Formative sites exist in substantial numbers in the hinterlands of the Puuc awaiting, like Paso del Macho, systematic survey.

Not only is Late Formative occupation widespread across the northern Maya lowlands, but there is also increasing ceramic evidence of regionalism developing between the eastern and western northern Maya lowlands, with the ceramics complexes of Cobá and Ek Balam beginning to contrast with those of Komchen and the west (Bey et al. 1998; Hernandez n.d.). Ceramic groups such as Chunhinta Black, Dzudzuquil, Tipikal, and Xanaba are found in greater abundance and diversity in the northwest part of the peninsula, while Carolina Bichrome, Dzilam, Huachinango Bichrome, and Valladolid Bichrome are more common and diverse in the east (Hernandez n.d.:41).

The Late Formative is also marked by the construction of major civic-ceremonial architecture. This is best seen at Komchen, where by 500 BC it had become a significant center, and by 300 BC it had come to dominate and incorporate other sites, including Dzibilchaltún (E. Andrews IV and Andrews V 1980; E. Andrews V 1981; Kurjack 1974). The architecture at Komchen included four major buildings (23F1, 24G1, 25O1, and 21J1) that were begun in the early Late Formative period (approximately 200–330 BC based on the C^{14} dates). Structure 21J1, built directly over a Middle Formative deposit, was a 39 m east-west platform supporting a pyramid more than 22 m across and preserved to a height of 2.8 m (Ringle 1999:194–95). In addition to these central structures, there also were a number of what Ringle calls "local temples." Several substantial platforms supporting secondary platforms were identified at Komchen, and Ringle would place Structure 603 of the Mirador group at Dzibilchaltún within this category (Ringle 1999:197).

Major civic-ceremonial architecture is also found at Late Formative Yaxuná. Important constructions include the 5E-19 group, a triadic group of mounds, much of the North Acropolis and the mounds it supports, and the two ceremonial buildings 6E-120 and 6E-53, associated with the East Acropolis. These last two structures are thought to have served as dance or performance platforms (Suhler 1996). The North Acropolis rises 26 m high and "appears from all indications to have been almost finished in its final form during the Late Formative" (Stanton 2000:536–37). According to Stanton, this was the high point of monumental construction at Yaxuná and the "achievements in monumental construction accomplished by later Maya of Yaxuná would never rival Late Formative developments" (Stanton 2000:537).

Although not as well defined as at Yaxuná, Komchen, and Dzibilchaltún, Late Formative civic-ceremonial architecture is also known from Ek Balam (FT-27 and GS-15), Kiuic (Yaxché group), Paso del Macho, Xocnaceh, many of the Costa Maya sites, and X-Huyub and Kax-ek in the Ek Balam region. The overall impression is that by the Late Formative, civic-ceremonial architecture was common and public labor investment was significant. The evidence also indicates that there was a great diversity in the type of structures and monuments constructed by this time.

Despite this evidence of growth, the Late Formative is also characterized by site abandonment in some areas, such as in the northwest area where after the Middle Formative period, overall site density drops until the Late Classic period (Robles Castellanos in press).

In other cases, as mentioned, Late Formative construction seems to be limited to renovation rather than major construction. The picture that is emerging is of a dynamic and complex landscape marked by regional variation in a number of dimensions.

A number of models have recently been developed to help explain the rise of complexity during the Late Formative in the north. In the early 1990s, Dunning (1992) summarized the largely processual thinking at that time, focusing on the idea that cultural complexity in the northern lowlands was driven by a set of pressures. These pressures include environmental and social circumscription, including competition for, and demand of, salt. He notes that Ball (1977) saw the northern lowlands as a circumscribed environment surrounded by oceans on three sides and rising population centers on the south. This led to territorial competition among the growing northern centers. It was this competition that was the prime mover leading to "political expansion and consolidation in conjunction with organizational innovation and development" (Ball 1977:171). The fact that the north coast of the Yucatán is one of the largest salt-producing areas in Mesoamerica and that many of the large Late Formative settlements such as Chunchucmil, Dzibilchaltún, Dzilam, Komchen, Tzeme, and Ucí are found within 50 km of the coast and run parallel to the northern salt mines, serve as evidence that salt was an economic force that played a significant role in the evolution of complexity in the north (Dunning 1992:64).

Despite the population growth associated with the rise of complexity in the Late Formative, there is little direct evidence for either population or resource pressure. The role of salt seems more compelling, especially when one considers the early evidence for complexity found by the Costa Maya Project in the Middle Formative along the coast (Robles Castellanos and Andrews 2003). However, how well does salt explain the growth of complexity in more inland areas such as the Ek Balam region or Yaxuná or the increasing evidence of Late Formative occupation in the Puuc region and along its northern boundaries?

Dunning also considered the role of religion as part of the model, suggesting that it served as a nucleating force that legitimized a growing elite (1992:64). They used the power and knowledge of an agrarian-based religion for organizing the ceremonial, administrative, and residential construction that appears in the new urban centers. More recently, Ringle (1999) has further explored the role of religion and ideology in the rise of social complexity in the northern Maya lowlands during the late Middle Formative and the Late Formative. He sees trade and specialization as playing minimum roles in this process. Except for Komchen, "which grew as the result of nearby concentrations of otherwise scarce resources, land and labor were the bases of wealth in most lowland Formative centers, wealth correlating directly with the ability to command labor and tribute" (Ringle 1999:189). The main goal of emerging elites was to successfully recruit and retain migrants. The emerging elites in the north developed a hierarchical architectural structure in the Late Formative with centralized platforms and pyramids, local or minor temples, and, in some cases, the placement of platforms next to local temples.

Ringle suggests that the adaptation of the northern Maya lowlands was unique: "local temples may have been prominent in domestic organization because organizational solutions developed elsewhere in Mesoamerica were inapplicable or undesirable" (Ringle 1999:197). He argues that economics did not mark differences in the north at this time, nor is there evidence that ceramics marking ethnic or residential identity were well developed.

The emerging hierarchy in the Late Formative used residential reorganization and ceremonial construction as a way to deal with issues developing due to increasing population levels, and perhaps dwindling availability of land, that demanded some concentration of authority. However, at the same time, Ringle sees recruitment and retention as issues that elites addressed through religion and ideology. They used ideology not to oppress but to integrate, creating centers as places of pilgrimage and ritual procession. This would explain the appearance of a *sacbe* (road) at Komchen during this period, and at the nearby site of Tamanche. Ritual architecture, focused on the center and the local, served to mediate between hierarchy and commoner of the emerging ruling lineages. He suggests that these centers served as places where religious cults were sponsored and supported. These cults provided a framework for the emergence of big men that did not involve hegemony or politically administrative structures as the prime movers (Ringle 1999:211). He sees the ideas associated with the emergence of complexity in the northern lowlands as having less to do with "legitimization than with providing a comprehensive view of society. I suggest that these metaphors were primarily employed during the Formative to define the relations between segments and paramounts. Hence they played a key role in recruitment and organization of the growing population" (Ringle 1999:214).

Ringle's model offers an alternative to hegemonic or politically administrative models dependent on resource control and management as the driving forces. Ringle (2005) is now adjusting his model to accommodate the rapidly changing views of the Formative, such as regards social complexity and ceramic differentiation. He no longer argues that prior to the Late Formative period, ritual was primarily done at the domestic level with little evidence of centralized ritual activity, nor that ballcourts are the next step in his emerging hierarchy as both are now found by the second half of the Middle Formative period. It will be interesting, given his ideas on the Middle Formative period, how Ringle synthesizes the Middle and Late Formative data and how he differentiates the social complexity of the Middle and Late Formative. There are clearly changes taking place between these two time periods, but the differences are now less clear. One thing that may be an important clue to the direction of culture change is the increasing regionalism found in ceramic complexes during the Late Formative.

One problem in evaluating these models is that the nature of Late Formative elite life in the north is still poorly understood. We lack major burials or tombs of elites, and there is very limited iconographic evidence from this time. The use of the *mat* symbols on Late Formative vessels at Ek Balam (Bey et al. 1998; Bond et al. 2003; Vargas de la Peña and Castillo Borges 1999) and the sculpted figure at Loltún Cave suggest that the idea of kingship and its related trappings became part of the stratification system at some point in the Late Formative. The widespread use of the step-fret motif on Late Formative/Early Classic Bichrome ceramics—such as Huachinango Incised-dichrome, Carolina Bichrome, and Dzilam Verde Incised-dichrome—is also thought to be associated with the evolution of social stratification in the north (Bond-Freeman et al. 2003).

Stanton thinks that by the Late Formative period, elite factionalism had developed within major northern centers such as Yaxuná, and that these factions were attempting to "establish social relationships with their peers across the Maya lowlands and possibly beyond, in order to gain access to prestige items in their wealth finance based economy" (Stanton 2000:577–78). He argues that the clustering of household structures at Yaxuná and Dzibilchaltún represent Late Formative factionalism based on kinship. He also finds the patterning of Late Formative monumental architecture at Yaxuná to be the result of competition among elite factions. He therefore sees elite power based on various elite factions focusing their energy on efforts to dominate and control centers. This model of elite life stands in contrast to Ringle's ideas that complexity grows out of cooperation and integration as opposed to competition.

The Early Classic

A little more than a decade ago, Dunning wrote that "the second half of the Late Formative period was marked by a wave of population loss and urban abandonment that spread south across the northern lowlands" (Dunning 1992:65). Until recently, this view of the Late Formative/Early Classic transition was widely held, representing a major misconception in our understanding of northern Maya Lowland archaeology. As mentioned earlier, Dzibilchaltún and Komchen do show major demographic declines at the end of the Formative period (E. Andrews V 1981, 1988; E. Andrews IV and Andrews V 1980). Their declines, coupled with the limited knowledge we had of the Early Classic in the north, led Maya archaeologists, with some exceptions (Ball 1977), to generalize these events as representative of the entire northern Maya lowlands between AD 100–700 (see Lincoln 1985:55). The north was not thought to recover until the rise of the Late and Terminal Classic states.

There is no doubt that just as in the southern Maya lowlands, the end of the Late Formative saw major population disruptions and site abandonment in the north; however, amidst this disruption was continuity, new growth, and increasing social complexity. Among the most significant Early Classic developments was the appearance of towns and cities constructing civic-ceremonial structures in the so-called Megalithic style, which was characterized by the use of large roughly cut stone masonry typically covered with a thick layer of stucco and modeled stucco iconography. Jennifer Mathews has scrutinized the characteristics, development, and extent of this architectural style (1998; see also Mathews and Maldonado Cárdenas, this vol.; Taube 1995). Long known from the sites of Izamal (Kurjack 2003; Lincoln 1980) and Aké (Maldonado 1980, 1989), the Megalithic style's relation and importance to the Early Classic in the northern Maya lowlands is only now being fully appreciated. Megalithic-style architecture defines the monumental architecture of the northern Maya lowlands during the terminal Late Formative and Early Classic periods, representing a regional expression of Maya culture in the north.

A wide range of structures was built using this architectural style, and Early Classic communities utilizing the Megalithic style appear across the northern plains, as well as in the Puuc and to some extent at sites along the admittedly amorphous border of the northern lowlands, such as Oxkintok, Cobá, and Yaxuná (see table 5.1, this vol. for a complete list of known Megalithic sites). The greatest concentration of these communities and probably the largest "pure Megalithic style" centers were located across the northern plains, from the Yalahau region in the east across to Izamal and Aké to the west (see fig. 5.1, this vol., for a map of the distribution of Megalithic sites).

The number of large early Classic Megalithic constructions in the north is truly impressive and includes such structures as the Kinich Kak Moo platform at Izamal (36 m high, 200 m × 200 m across with a 100 riser Megalithic stairway and a 15 m high Megalithic-style pyramid); Structure 2 at Ox Mul, Quintana Roo (75 m by 45 m) (Glover and Esteban-Amador 2002, 2005); the 110 m long, 80 m wide, 9 m high Structure 3 at Ucí, Yucatán (Maldonado 1980, 1995); the 18,000 m² platform at Victoria, Quintana Roo, supporting 10 structures (Glover and Esteban-Amador 2004); and the 8 m high pyramid at Yaxhom (Mathews and Maldonado Cárdenas, this vol.). Many of these Early Classic Megalithic centers were equal in size and scale of construction to those appearing at the same time in the southern lowlands, indicating a high degree of social and political complexity.

There is still much to be worked out about the nature of Early Classic Maya culture in the northern lowlands, including even the basic chronological history of the Megalithic style. Although Mathews and Maldonado Cárdenas (this vol.) favor dates of 150 BC–AD 400 for the style, buildings at Ek Balam (GT-10) and at Chac II in the Puuc region suggest the style may have continued later. Smyth's excavations of the Great Pyramid at Chac II in the Puuc indicate the initial construction phase was dated to AD 400. It was followed by a "foreign style" construction phase, then an Early Puuc enlargement, and finally a Megalithic style. On this basis, Smyth pushes this particular example of the Megalithic style into the Late Classic/Terminal Classic (Smyth and Ortegón Zapata, this vol.). Late Classic sherds were also recovered from test pits placed atop GT-10 at Ek Balam.

Although it seems to have its greatest expression in the northern plains, Megalithic construction extends down along the southern boundaries of the northern Maya lowlands, appearing to varying degrees at such sites as Oxkintok, Chac II, and Yaxuná. These sites and others like Chunchucmil are primarily associated with different Early Classic architectural styles affiliated with the southern lowlands.

Oxkintok is a major center by the Early Classic period, where "Early Oxkintok" (Varela Torrecilla and Braswell 2003) architecture is associated with "four lintels dated to the fifth century and Stela 4, a monument stylistically related to contemporaneous southern Maya Lowland sculpture" (Stanton 2000:573). This suite of cultural traits connects this section of the northern lowlands more directly with

Petén and southern Maya lowland developments. There is, however, Megalithic stonework found at Oxkintok, which leads Jennifer Mathews to conclude that "while none of the structures at Oxkintok are constructed in the purely Megalithic style . . . some sort of sharing of ideas was occurring during the Late Formative and Early Classic periods in this western end of the interaction sphere" (1998:152).

At Yaxuná, Stanton (2000:568) proposes that what took place was an expansion of the northern Megalithic culture, which he considers to reflect a possible hegemony emanating from Izamal into a site which had heretofore had strong ties to the Petén. He sees the appearance of a Megalithic platform combined with a lack of Petén ceramics as evidence supporting this hypothesis and dates the event to the Late Formative/Early Classic transition. By AD 250–400 Yaxuná frees itself from the Megalithic hegemony and, based on architecture and the appearance of "polychrome ceramics with strong modal ties to the Petén" (Stanton 2000:561), reasserts its ties to its southern lowland allies.

Chunchucmil, located to the west of Oxkintok near the coast, is argued to have emerged as a major northern specialized trading center in the Early Classic (Blackmore and Ardren 2001; Stanton et al. 2000). Although occupied since the Middle Formative, the site reached its massive size in the Early Classic (see Hutson et al., this vol.). At that time Chunchucmil covered as much as 25 km^2, had a population estimated to be at least 30,000, and had some of the densest occupation in the entire Maya world. Interestingly, there is no evidence of Megalithic architecture, and the ceramics suggest that most of this growth took place rapidly during the late Early Classic. These facts, combined with what is known from Yaxuná and Oxkintok, suggest that the Megalithic style was not as prevalent at the edges of the northern Maya lowlands as it was further north, nor was it part of the dramatic Late Early Classic growth associated with these centers.

The importance of long-distance trade in the Early Classic is recognized not only at Chunchucmil. Xcambó (Sierra Sosa 1999, 2001:27) was another major Early Classic trading port located on the north coast of the peninsula. A number of structures at the relatively small site combine Megalithic- and Petén-style construction and, based on the associated ceramics, are dated to the Early Classic (AD 250–600) (Sierra Sosa 1999). The site of Xcambó became a mercantile center involved in the production and distribution of salt during the Early Classic. Xcambó is notable for the significant amount of trade goods recovered within the site, including ceramics from Cobá and the east coast as well as the Petén, especially polychromes (Sierra Sosa 1999).

The last two decades of fieldwork have provided clear evidence for

the widespread existence of an Early Classic culture in the northern
Maya lowlands. It was marked by regional diversity, as well as by increasing complexity and scale of construction. Sites such as El Naranjal, Aké, Izamal, and probably Kantunilkin in the north, and Oxkintok,
Chunchucmil, Chac II, and Yaxuná in the south represent the wide
range of major centers in the northern lowlands. Future research will
no doubt establish that others, such as Yaxhom in the Puuc, were also
large Maya centers during the Early Classic. Work at Chunchucmil and
the northern coastal site of Xcambó reveal that the north was involved
in significant long-distance trade with both the southern lowlands and
the rest of Mesoamerica.

It is also important to note that northern Maya lowland distribution patterns vary from region to region. For example, in the Ek Balam
region, Early Classic sherds are found across much of the central area
of the site of Ek Balam, though they are surprisingly rare in our rural
collections. On the other hand, Kepecs found Early Classic material
at 90 percent of the sites she surveyed in the adjacent Chikinchel region (Kepecs 1998:124). Not only should we expect regional variation in site size and rural populations, but also in architecture. I doubt
that all buildings built in the northern plains, for example during the
Early Classic, are Megalithic. Monumental construction should be expected in other styles, such as the mixture of Megalithic and Petén
styles found at Xcambó.

Explanatory models for the evolution of Early Classic Maya culture
in the north are poorly developed. Stanton hypothesizes the existence
of an Early Classic hegemony emanating from Izamal (2000:568). Jennifer Mathews (1998) alternatively argues the distribution of the Megalithic architecture associated with the end of the Late Formative and
the beginning of the Early Classic represents the archaeological remains of an interaction sphere. This interaction sphere evolved as the
elites in the northern Maya lowlands created a regional information
exchange network. This exchange network developed on a local and
regional scale among the elites, out of a demand for the exchange of
scarce or critical resources (J. Mathews 1998:5–6). Elites controlled,
through their regional network, the distribution of raw materials and
finished products but not the means of production. In her model, it
is not the intensification of trade in exotic goods that drives complexity but the fact that a set of local economies have merged into a
regional one in which regional exchange becomes essential for maintaining local economic integration. Elite power grows in this model
when elites are able to monopolize the interaction occurring between
the regional and local networks (J. Mathews 1998:6).

Despite the growth of a possible interaction sphere, the picture presented by Early Classic ceramics is of increasing differentiation. According to the most recent ceramic synthesis, the regionalism that began in the Late Formative develops into at least five regional ceramics spheres by the Early Classic (AD 250–600) (Gallareta Ceballos and Jimenez Alvarez n.d.). These spheres include one in the north center of the peninsula, one in the greater Puuc area, one centered around Yaxuná, one centered on the Ek Balam region, and one extending along the southern east coast. These spheres also include differing trade wares and influences, indicating another level of interaction defining them. Furthermore, it now appears these Early Classic spheres show, to a surprising degree, continuity into the Late Classic (Robles Castellanos n.d.), supporting the idea of a largely in situ ceramic evolution beginning in the Formative.

Thus, our present understanding of the Maya world in the north supports a case for strong in situ evolution beginning no later than the Middle Formative and continuing through the Early Classic into the Late Classic. The Cehpech ceramic tradition (AD 700–1050), with its emphasis on slate wares, is now recognized as having evolved in situ from certain Early Classic types found in these complexes. Our work at Ek Balam provided a solid line of development from pre-Cehpech to Early Classic slates to the Late Classic forms of slate wares diagnostic of the Cehpech sphere (Bey et al. 1998).

What remains lacking is the existence of a significant body of iconography or glyphic texts associated with these Early Classic centers in the north. In some ways, the northern Early Classic seems to have more continuity with the Late Formative than in the south. The impression of northern Maya culture has long been that it has a dearth of texts compared to the south. This has been thought to be a result of the relative lack of an Early Classic occupation in the north. Traditional thinking was that when the southern Classic culture reached the north it did so at Oxkintok, Cobá, and later at northern sites like Dzibilchaltún and Ek Balam, where we see evidence of it. Despite the fact that this model is incorrect, nonetheless it still remains that the overall level of literacy among the northern Maya during the Classic period appears to be significantly more limited. As discussed earlier, there are no sets of monuments providing long dynastic sequences for the Classic Maya in the northern Maya lowlands nor did potters produce polychrome vases with texts like in the south. This lack of literacy is seen both in the Early Classic and in the Late/Terminal Classic Puuc tradition, where texts were evident but rare compared to the southern Maya lowlands.

Based on our new understanding, the difference in literacy is the outcome of a long history of regional development of northern Maya lowland society. The Early Classic Maya in the north were not country bumpkins living in the backwaters of the Maya world but instead a culturally distinct group that took a different trajectory to some extent from that of the south. The limited number of texts and polychromes in the north is not a result of the area being culturally backward but the result of an indigenous and still not-fully-understood regional cultural tradition.

Discussion and Conclusions

The results of the last quarter century of research have dramatically altered our understanding of the prehistory of the northern Maya lowlands. It is no longer useful to consider northern Maya prehistory as a pale reflection of the south or a mere reaction to southern political and historical events. It is an area of cultural development with its own history and identity that begins in the Middle Formative and continues through the Late Postclassic. The scale of occupation and the degree of cultural complexity found in the northern lowlands during the Middle Formative has caused us to readjust our models of the nature of the initial settlement of the region. The Late Formative is now understood as a robust and complex expression of a regional Maya culture that follows a path to complexity that is connected to, yet independent of, that of the southern lowlands and that appears to necessitate explanations that are, to a degree, unique to the region. Perhaps most exciting is the growing understanding of the nature and scale of Early Classic Maya society in the north. It is here that we see the roots of Classic culture in the north providing us with the information to now recognize that Late and Terminal Classic culture are largely in situ phenomena and not simply the last gasp of the southern Classic Maya.

The changes in our understanding are due to the impressive amount of research that has been carried out in the north. Not only do we know much more about sites that have long been studied, but now entire areas such as the northeastern portion of the peninsula, which was unknown twenty years ago, are recognized as playing important roles in the developments of the region. Other areas like the Puuc, long considered a single-component region, are now being reevaluated as having long sequences of development that must be explored in order to understand the prominent Terminal Classic cities of the zone.

Our need for continued research in the north remains great. We have very little evidence of the Late Formative elite other than the monumental architecture, and little evidence of non-elite life at all.

We have an enormous call for projects focusing on the Early Classic. Recent projects such as those at Chunchucmil and Chac II are providing important evidence for better understanding the Early Classic in the north, and it is my belief that the Early Classic will continue to become a more important area of research in the coming years.

While this chapter covers a great deal, space does not permit me to fully explore the information available on the time periods covered, such as the Early Classic occupations at Cobá, Chunchucmil, Oxkintok, Xcambó, and Chac. I avoid discussing the nature and extent of interaction between Teotihuacan at these and other northern sites (see Varela Torrecilla and Braswell 2003), as well as the transformations that take place in the north between AD 450–700. I also do not address the time periods where the majority of research has been carried out over the last twenty-five years, the Late and Terminal Classic and the Postclassic. The goal here has been to provide a different context for considering these later developments, many of which are discussed in this volume.

Part 2 Making a Living

2

From Swidden to Swamps
The Study of Ancient Maya Agriculture
Bethany A. Morrison

Most agriculture practiced by modern Maya populations is known as swidden, or slash-and-burn, cultivation. This process includes the clear-cutting of forest and the burning of the cut vegetation. Ash from the fires returns nutrients to the soil, into which can be planted seeds, usually the triad of corn, beans, and squash. This labor-extensive system produces relatively low yields and requires high amounts of acreage per individual to be fed. Since the 1970s, archaeologists working in the Maya lowlands have found increasing evidence for pre-contact population levels considerably higher than those of today, prompting the question: how did the ancient Maya produce enough food to support those populations, including many non-food-producing groups such as craftsmen, priests, and merchants?

Research by the Yalahau Regional Human Ecology Project has attempted to address the problem of how ancient communities in the northern Maya lowlands sustained themselves (Andersen 2001; Fedick 1996b, 1996c, 1998b; Fedick and Hovey 1995; Fedick and Taube 1995; Fedick et al. 2000; Morrison 2000). Since 1993, this project has been based in an area of the northeast corner of the Yucatán Peninsula referred to as the Yalahau region. Within this region is a series of wetlands known as the Holbox Fracture Zone. Although these wetlands differ significantly from those of the southern lowlands in both hydrology and ecology, reports of anthropogenic features within one of the wetlands led project members to believe that it might also have been used at one time for some form of intensive cultivation. A five-year process of creating hypotheses and testing them, creating new, alternative hypotheses and testing them, and so on, has modified our ideas about this wetland and its purpose considerably. The project described below also highlights the way in which archaeologists often must combine several academic disciplines and/or methodological approaches to answer a single question. In this case, settlement pattern analysis, chemical ecology, malacology, topographic survey, and other techniques are utilized.

The Swidden Thesis

From the earliest accounts written by explorers such as John Lloyd Stephens (1843), the Maya landscape was perceived as a uniform sea of dense vegetation and shallow, rocky soils. Similarly, archaeologists and soil scientists alike have characterized tropical soils as poor in nutrients and unable to support intensive cultivation (cf. L. Alexander and Cady 1962; Chang 1968; Fedick 1996c:1–2; Karmack 1962; McNeil 1964; Sivarajasingnam et al. 1962). In 1946, Sylvanus Morley wrote that swidden agriculture "is practically the only system of agriculture practiced in the American wet tropics even today, and indeed is the only method available" and "the modern Maya method of raising corn is the same as it has been for the past three thousand years or more" (1946:141). This swidden thesis became doctrine, shaping archaeologists' views of Maya cultural development and social organization into a paradigm of a uniform environment, extensive agriculture, and resultant low population levels (Fedick 1996b). Limited yields from swidden agriculture were even hypothesized to have caused the collapse of Classic Maya civilization. This paradigm lasted into the 1970s, despite much evidence to the contrary (e.g., Bronson 1966; Gann 1925; Lundell 1933; Ower 1927; Palerm and Wolf 1957; Puleston 1968; Ricketson and Ricketson 1937; Schufeldt 1950).

Alternatives to Swidden Agriculture

In *The Myth of the Milpa: Agricultural Expansion in the Maya Lowlands*, Hammond (1978) marks the end for the swidden thesis. With the introduction by Gordon Willey of settlement pattern archaeology into the Maya area (Willey et al. 1965), detailed surveys of hinterland areas brought to light two significant pieces of evidence that stood in direct opposition to existing beliefs. First, archaeologists were realizing that there were higher ancient population levels than could realistically have been supported by swidden agriculture. Second, they were coming across vestiges of intensive field modifications such as terraced hillsides (Turner 1974). This mounting evidence meant that archaeology could no longer cling to the simplistic ideas of the swidden hypothesis.

Since then, subsistence has become part of the research agenda of nearly every archaeological project conducted in the Maya area. Although tropical soil conditions often limit the preservation of botanical remains, evidence (such as pollen and carbonized seeds) of several plant foods has been uncovered at various sites throughout the region. As predicted by the otherwise limited swidden thesis, corn was a substantial portion of the ancient Maya diet, which was complemented by beans and squash. Other plant remains that have been recovered include chili peppers and manioc, a root crop that had been suggested as early as 1966 (by Bronson) as a potential alternative to corn and a major source of calories in a tropical environment. Additionally, a wide variety of tree fruits have been recovered, including cacao, nance, avocado, mamey, calabash, guava, and papaya (see Lentz 1999 for an excellent review of botanical remains recovered from Maya sites). However, the ramon nut, also proposed as an alternative to corn (Puleston 1968), has not been significantly recovered from archaeological deposits.

Ethnographic research has highlighted the use by modern Maya of home gardens, or planting beds and orchards within the walls of house compounds, which produce herbs, fruits, and vegetables, providing much of the variety in the modern Maya diet (see Heidelberg and Rissolo, this vol.). These home gardens were undoubtedly also grown within ancient Maya compounds, and while they were likely to have provided much of a household's sustenance, they are not thought to have produced significant surpluses. However, while the traditional view of Maya agriculture has been of a system of outlying *milpa* (corn) fields complemented by house gardens, excavations at the Pompeii-like site of El Cerén in El Salvador have revealed a stand of corn growing immediately adjacent to a residence, indicating a more intensive, localized pattern of production (Sheets 1992).

Another form of intensified farming efforts by the ancient Maya is the use of terraced fields (Beach and Dunning 1995; Turner 1974). Terracing is a labor-intensive modification to the landscape, which affects the temperature, hydrology, and stability of sloped fields, often allowing agriculture in areas where it would otherwise be impossible or relatively unproductive. Evidence of terracing greatly expands the agricultural resources potentially available to the ancient Maya and increases estimations of crop yields. This evidence is limited to areas of well-drained slopes, such as are found in the southern Maya lowlands, and are not available farther to the north where the terrain is generally very flat.

Other forms of intensification have been identified across the Maya area, revealing a system by which local environments were manipulated on an individual basis to maximize agricultural potential (Fedick 1998b). As will be discussed below, the use of wetland fields has been the focus of considerable investigation and has revealed considerable variation in the techniques used, including terracing, agroforestry, and hydro-management methods, such as drained and raised fields in wetland environments.

Wetland Agriculture

In 1972, a report by Alfred Siemens and Dennis Puleston of patterned ground in riverine-associated wetlands of Campeche, Mexico suggested for the first time that intensive wetland cultivation might have played a role in sustaining prehistoric Maya populations. This report, as well as an edited volume by Turner and Harrison (1983) regarding Pulltrouser Swamp of Belize and another by Harrison and Turner (1978) entitled *Prehispanic Maya Agriculture*, utilized evidence of raised fields to challenge the notion that the ancient Maya had relied solely on swidden agriculture. These new ideas thereby brought hypotheses on Maya subsistence into better accord with the growing evidence for large prehistoric populations (see also Hammond 1978).

These and more recent investigations have concentrated on three main types of wetland environments in the Maya area: the coastal mangrove swamps, the riverine and permanent swamps of the central and southern Maya lowlands, and seasonally inundated areas including *bajos* (depressions) and the wetlands of the Holbox Fracture Zone in the Yalahau region. Coastal swamps, however, are not of consideration concerning the cultivation of domestic crops, given their brackish nature.

Investigations of prehistoric wetland agriculture have concentrated primarily on the central and southern Maya lowlands. Studies of permanently inundated wetlands include reports from Campeche (Siemens and Puleston 1972) and northern Belize (Harrison 1990, 1996; McAnany 1992; Pohl 1990; Pyburn 1989, 1996; Turner and Harrison 1983). These swamps are either permanently inundated or flooded annually by silt-bearing rivers and contain a rich supply of organic mucks that are a good source of nutrients for cultivated crops. Permanent planting beds were created within these swamps either by channeling away water (drained fields) or by building up platforms (raised fields) similar to the famous *chinampas* of central Mexico.

Studies of bajos include work in Petén, Guatemala (Culbert et al. 1990), northern Belize (Kunen 2004), and southern Quintana Roo (Gleissman et al. 1983). However, bajos and the Yalahau wetlands of the northern lowlands present a distinctly different wetland environment. These are typically seasonally inundated areas, offering a much less stable environment for cultivation. The Yalahau wetlands are also notable for containing shallow, clay-like soils over bedrock. The shallow nature of these soils and their low organic content preclude their use in the construction of raised fields like those in the muck-rich swamps of the central and southern lowlands.

Studies of agriculture in seasonal wetlands have included work at the raised fields of Bajo de Morocoy in southern Quintana Roo (Gleissman et al. 1983), a bajo near Rio Azul in northern Petén (Culbert et al. 1990), a channeled bajo at La Justa in central Petén (Kunen et al. 2000), and the Yalahau wetlands of El Naranjal and El Edén (Andersen 2001; Fedick 1998a, 1998b; Fedick and Taube 1995). The Yalahau wetlands, particularly the El Edén wetland, were managed with the use of dikes and check dams (Andersen 2001; Fedick 1998a). In the Yalahau cases, the use of different technology to manage wetlands of a very different nature than those of the southern lowlands may also mean that they were used at a different time, that they were used to grow different crops, or that they played a different role in the developmental history of Maya civilization.

Problems/Questions

Although our understanding of ancient Maya subsistence has come a long way from the unsubstantiated swidden thesis of old, the northern Maya lowlands remain something of an enigma. Without the undulating terrain or permanent swamps common farther to the south, the northern lowlands could not be exploited through the use of terracing, or raised or drained fields. The question of how northern Maya lowlands populations sustained themselves remains unanswered. It is assumed that large pre-contact populations required some system of intensification to produce an adequate food supply, but what specific means of production were used is still uncertain. Work by members of the Yalahau Regional Human Ecology Project is beginning to provide some answers.

Current Research

The Yalahau Wetlands

The area of wetlands, referred to as the Yalahau region, has historically been inaccessible and sparsely populated (see Fedick 1996b). The extremely thin soils above karstic limestone bedrock make modern agriculture difficult and keep production yields low. The wetlands themselves, however, are an unusual source of fresh water in the northern peninsula, and recent discoveries of a complex array of prehistoric Maya sites surrounding these wetlands suggest that the area was once much more productive and able to support a substantial population (Fedick & Taube 1995; Glover and Esteban Amador 2005; Morrison 2000).

The Yalahau wetlands, particularly one located on the El Edén Ecological Reserve, were altered by the ancient Maya to increase or sustain their usefulness. This was done through the construction of earthen check dams built over a skeleton of upturned limestone slabs. These check dams have subsequently eroded, but the slabs remain as extant rock alignments that can be mapped with relative ease if one is willing to conduct a systematic survey in the soggy wetland terrain (fig. 2.1). The construction of these check dams would have required a significant investment of labor, indicating that the wetlands were central to the ancient Yalahau economy (Fedick 2003; Fedick et al. 2000). Many potential uses have been suggested for these wetlands (see Fedick 1998b, 2003; Fedick et al. 2000; Morrison 2000). For example, corn or other crops could have been planted on the margins of the wetlands during the otherwise unproductive dry season. Additionally, alterations to the wetlands may have served to improve the environment for naturally occurring and economically useful flora and fauna such as palms, fruit trees, fish, turtles, or the edible apple snail.

Another hypothesized use of the altered wetlands was first suggested during a conversation with Ana Luisa Anaya, a chemical ecologist who is a member of the multinational, multidisciplinary team studying the El Edén Ecological Reserve. This idea was that the wetland may have been used for the production of a natural fertilizer. Every year, during the rainy season, the Yalahau wetlands become covered with a green carpet of algae and other organisms called periphyton, among which thrives a community of small freshwater mollusks, such as snails, bivalves, and limpets. This periphyton has been found to be very high in nitrogen and phosphorous, higher even than modern chemical fertilizers (Anaya et al. 1997; Palacios-Mayorga et al. 2003). Periphyton or periphyton-enriched wetland soils could have been collected for use in upland fields, significantly increasing their yields in an area otherwise plagued with poor soil quality. The study reported here examines soil samples collected from an ancient Maya settlement associated with the wetland at El Edén, in hopes of uncovering evidence for the transport of periphyton or wetland soils out of the wetland and into areas of either milpa or homegarden cultivation.

Fig. 2.1 Example of a rock alignment in the El Edén wetland. This feature, crossing the north end of the wetland, measures 700 meters in length. (Photo by Scott L. Fedick, 1995.)

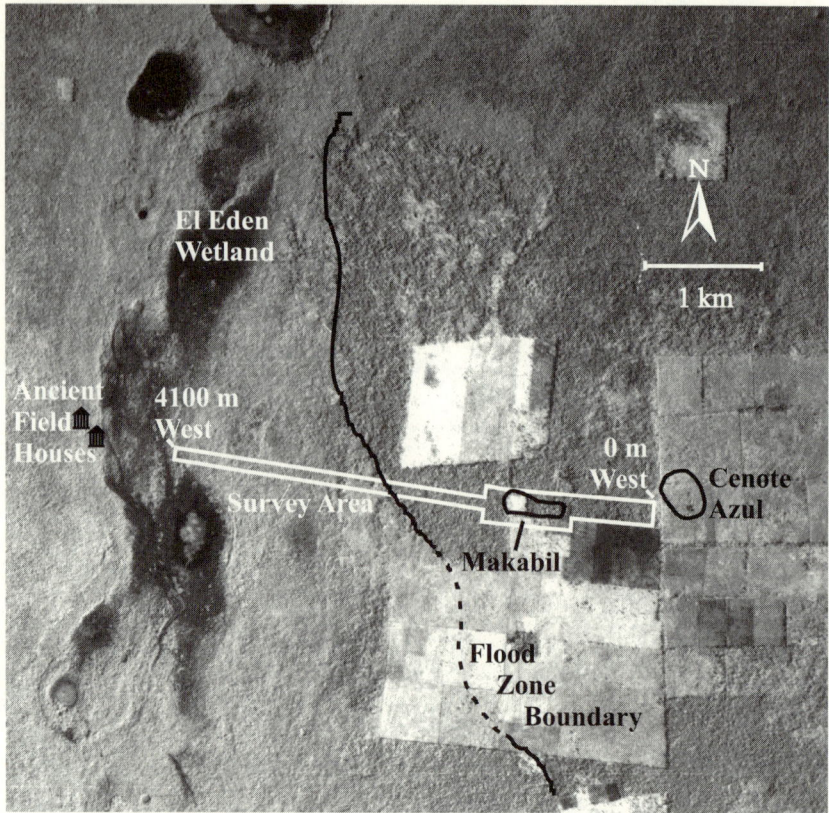

Fig. 2.2 An aerial photograph of the El Edén wetland, indicating the locations of Cenote Azul, Makabil, the survey area, and the approximated eastern limit of the wetland's flood zone. Measurements along the survey baseline were recorded in meters west of Cenote Azul. (Image created by Bethany A. Morrison.)

The Site of Makabil

The research reported here began as a settlement survey. The intention was to discern the relationship, if any, between the El Edén wetland and a small elite settlement, called Cenote Azul, located 4 km to the east. A 100 m wide transect was surveyed between these two points, revealing an additional, discrete settlement located between 800 and 1,400 m west of Cenote Azul (about 3 km east of the wetland). This site was given the name Makabil (fig. 2.2).

Ceramics recovered from Makabil place occupation during the Late Formative period (ca. 100 BC–AD 350), a date consistent with the occu-

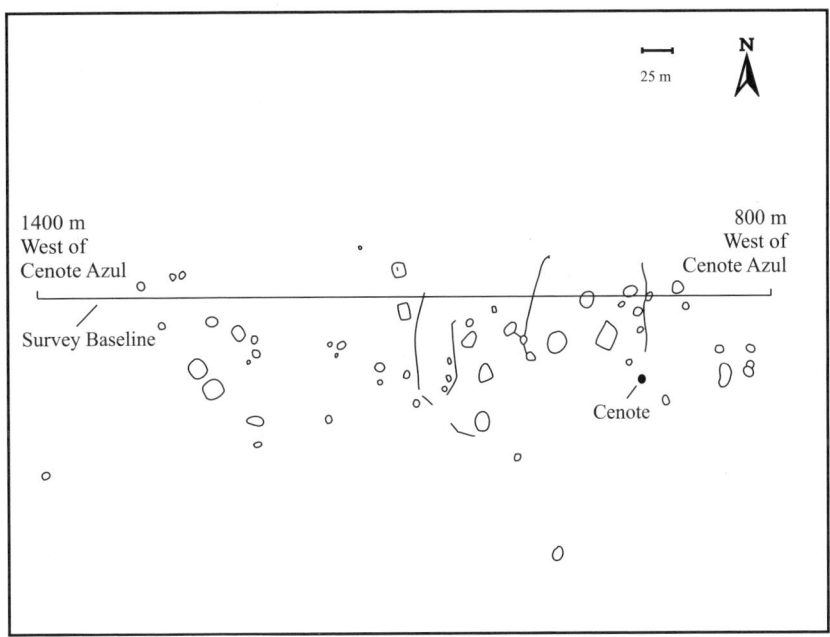

Fig. 2.3 The settlement of Makabil. Walls are interpreted as indicating the boundaries of ancient house lots or *solares*. (Map created by Bethany A. Morrison.)

pation of other sites in the Yalahau region (Fedick and Taube 1995; J. Mathews 1998; Morrison 2000). Makabil consists of sixty-two structures of limestone rubble and roughly shaped blocks that each probably supported one or more perishable structures. Most of the platforms are less than 100 m³ in volume and likely served as outbuildings such as kitchens and storage facilities. There are fourteen platforms at Makabil that clearly seem large enough to have supported residences, but there is no elite or obviously monumental architecture at the site. Makabil is an example of a hinterland settlement, probably primarily reliant on farming to support its inhabitants.

The Makabil settlement is divided into five sections by a series of low stone walls that tend to run parallel to each other in a north-south alignment, breaking up the settlement, which is distributed along an east-west axis (see fig. 2.3). Walls similar to these, surrounding and separating residential units at sites such as Cozumel (Sierra Sosa 1994), Cobá (Folan et al. 1983), Mayapán (Bullard 1962), and Playa del Carmen (Silva Rhoads and Hernandez 1991), have been likened to modern-day *solares*, or houselots (Herrera Castro 1994; Herrera Castro et al. 1993) and interpreted as household boundary markers and evidence of ancient home gardens (see Goñi Motilla 1993; see also Heidelberg and Rissolo, this vol.). This would suggest that whatever the primary farming technique used by the residents of Makabil was, it was supplemented by products (herbs, vegetables, fruits) produced closer to home in these walled-off gardens. So, the settlement survey itself provided some insight into prehistoric subsistence in the area. Upland milpas and/or these home gardens could have benefited from the application of nutrient-rich periphyton transported from the nearby wetlands.

Settlement around the El Edén Wetland

Additional roadside, milpa, and *cenote* (karstic sinkhole) surveys, combined with information provided by local informants, led to the discovery of several other ancient Maya sites in the vicinity of El Edén. These non-systematic survey efforts concentrated on areas east of the El Edén wetland and those immediately north and south of it, covering approximately 200 km². To date, thirteen sites have been identified within the reconnaissance area. The location of each site was recorded with a global positioning system (GPS).

Based on the estimated volume of largest structure, sites in the area can be divided into three sizes. Large sites, each with one or more structures of over 6,000 m³, include the site of Yax Meex to the north, then Xux, and then T'isil farthest south. Medium sites, generally including one elite residential compound and each containing one or more structures between 900 and 1,500 m³, include the sites of Cenote Azul, Carmelita, and S-95-8. Each of the large- and medium-sized sites is located adjacent to a large cenote or *nauahuela* (cenote with little water in it). Finally, there are seven sites known to date with no noticeably elite architecture, or in other words no structure over 450 m³. As these sites are very small and tend to be made up of low-lying foundation platforms, it is likely that many similar sites in the area remain undiscovered. The only one of these small sites that has been investigated in detail is Makabil, and it was found to be associated with a small, bell-shaped cenote.

Although a conclusive discussion of the settlement pattern of the El Edén area must await a systematic survey, initial interpretations can be made based on the data available at present. As concluded by Bell (1998), the large and medium sites are all located adjacent to a large cenote or nauahuela. Generally speaking, the large sites mark the eastern edges of the known settlement, farthest from the wetlands. The medium sites, smaller but still containing elite architecture, lie between the large sites and the wetlands. The smallest sites are scattered among the medium sites or are located between the medium sites and the wetlands.

The location of the large sites away from the wetlands (rather than, for example, in association with the large cenotes closest to the wetlands) suggests that direct physical control over access to wetland resources was not a priority of the local elite. The large sites may rather have been located to take advantage of some other resources. For example, the nauahuela at T'isil is today a source of wild vanilla; the unusual microclimate allowing the growth of this plant may have been what attracted the ancient Maya elite to the site (Mathews and Fedick 2000).

The three medium-sized sites may have been agricultural communities containing one or more large extended families whose farming successes had allowed the accumulation of some wealth and therefore some elevation in their status. Regarding the small sites, most of which were surely communities of subsistence farmers, it seems reasonable to assume that residents of the sites closest to the wetlands, particularly Makabil and Natsak'al, would have had the most direct physical (if not social) access to the wetlands and their resources or products. Although the exact relationship between these sites and the wetlands will remain unclear until we have a better understanding of how the wetlands were used, Makabil and Natsak'al may have been communities of agricultural or other laborers who specialized in the cultivation or other utilization of the wetlands.

It should also be noted here that systematic survey of the El Edén wetland by Fedick in 1998 identified two features on the western edge of the wetland. Excavations of these features by Morrison (2000) suggest they are field house foundations. These field houses would have facilitated bouts of intensive activity in the wetland by individuals living in the local settlements, each 3 km or more away.

With the knowledge that the Yalahau wetlands and the El Edén wetland, in particular, supported a hierarchy of settlements, it becomes all the more interesting to discern the nature in which those wetlands were being exploited. If periphyton had been extracted from the wetlands, is there any evidence for it in outlying areas that may have been milpas? Is it found within settlements? If so, is it associated with a specific level of the social hierarchy?

Topography of El Edén

In order to discern where periphyton might grow naturally versus where it would be found if someone had put it there, it was necessary to determine the maximum extent of the wetland during times of high water. The terrain of the Yalahau region, especially around El Edén, is remarkably flat. Due to the shallow relief, many of the area's wetlands have no clear margins, and inundation during the rainy season can cause them to increase greatly in size, flooding surrounding areas. In order to determine the extent of the flood zone associated with the wetland at El Edén, a detailed topographic study was conducted along the survey transect leading from the dry-season margin of the wet-land, through Makabil, and on to Cenote Azul (see fig. 2.2). As the topography of the Yalahau region consists of subtle ridges and depres-sions that run north-south, measurements taken along the east-west baseline would reflect a cross section of the local terrain (for a detailed discussion of the archaeological survey of this transect, see Morrison 2000).

At the time of the project, there was no point on or near the El Edén Ecological Reserve with a known elevation. As the project was con-cerned with elevations relative to flood levels, it was decided that mea-surements would be taken relative to the high-water mark from the 1995 rainy season, a point that is permanently recorded on a support piling of the El Edén research station structure. The American Meteo-rological Society (AMS) recorded the hurricane season of 1995 as the second most active season in 125 years, falling short only of 1933 for number of storms, while breaking the record for days of storm activity (AMS 1995). In view of this, the flood levels from 1995 are considered to be a good approximation of high-water levels for the area over time. Elevations were measured to the nearest centimeter with a transit and a stadia rod.

Remarkably, elevations along the baseline only vary by 4.3 m. About 300 m east of the dry-season margin of the wetland, at 3,900 m west, the bedrock rises to 1.2 m above the dry-season water level and 2.8 m below the 1995 flood level. Here, the vegetation changes from open savanna and *palo tinto* trees (spp. *Haematoxylon campechianum*) to denser, higher forest. During most years, this point probably marks the wet-season extent of the wetland. However, waters as high as those in 1995 could spread as much as another kilometer east to 2,800 m west, where a ridge was recorded 0.75 m above the 1995 flood levels. This estimation of the extent of the El Edén flood zone is supported by evi-dence visible in the modern terrain. Depressions west of this ridge are marked by a lack of vegetation and high densities of snail shells, indicating prolonged periods of high moisture or of inundation. Additionally, aerial photos of the reserve and surrounding areas indicate a change in vegetation types associated with this ridge, also suggesting a difference in soil quality and/or moisture levels. The boundary of the flood zone indicated in fig. 2.2 follows this change in vegetation, crossing the transect at the ridge.

It appears, therefore, that Makabil is located 1.5 km outside of the extreme flood zone of the El Edén wetland, secure from inundation and also well outside the natural habitat of periphyton and the aquatic species of mollusks that are part of that biotic community.

Searching for Periphyton at Makabil

Among the community of organisms that makes up periphyton are several species of small freshwater mollusks. These creatures are significant to this study for two reasons: (1) their shells, unlike other parts of the periphyton, will preserve archaeologically; and (2) they cannot survive naturally outside of the wetland environment. If periphyton, or the substrate immediately beneath it, had been transported to the settlement or other areas outside of the wetland for the purpose of soil augmentation, some of these shells should have been transported along with them. These shells, then, should be recoverable from soil samples collected in the treated areas.

Soil samples of 10 liters were collected at intervals along the baseline between the El Edén wetland and Cenote Azul at maximum intervals of 200 m. The soil samples were water-screened with 1 mm mesh, and all material retained in the screen was then dried and prepared for analysis. Mollusk shells were sorted out from the processed soil samples and then sent to Roberto Cózatl Manzano, a malacologist with expertise regarding the mollusks of the El Edén Ecological Reserve, for species identification and verification of either a terrestrial or freshwater niche for each specimen (Cózatl Manzano 1999; Morrison and Cózatl Manzano 2003).

Results

In all, six freshwater and seven terrestrial species of mollusk were identified to at least the family level (for a detailed discussion of the identification process and exact distribution of recovered mollusks, see Morrison 2000; Morrison and Cózatl Manzano 2003). As expected, frequencies of freshwater mollusks decline dramatically beyond the edge of the permanent section of the wetland (from hundreds per sample to a dozen or fewer) and drop off completely (to zero recovered freshwater mollusks) around the protective ridge marking the edge of the flood zone. Then a few specimens occur again, west of the ridge, outside of the flood zone, and outside of the natural habitat of the freshwater mollusks. Significantly, these occur between 800 m and 1,600 m west, precisely within the settlement of Makabil. Although the actual number of freshwater mollusks found at Makabil is small (only eleven), they clearly show the presence of transported wetland species.

The possibility remains, however, that the freshwater mollusks were transported by some natural means, perhaps by unusually high floodwaters. In such a situation, mollusks outside of their natural habitat ought to reflect a random distribution. A chi-square test was performed, indicating a 99.5 percent assurance of a non-random distribution of freshwater snails outside of the flood zone (between 0 and 2,500 m west of Cenote Azul). In other words, the mollusks, and presumably the periphyton in which they once lived, were put there intentionally.

Discussion and Conclusions

As the result of a simple conversation among colleagues at El Edén, including a little inductive brainstorming, what had begun as a basic settlement survey quickly turned into a much more complex project requiring information on topography, hydrology, and biology. It also turned into a project with potentially much greater implications.

Freshwater mollusks recovered from soil deposits at Makabil are exciting evidence for a previously unrecognized form of intensive agriculture among the ancient Maya. The support of periphyton communities was likely one purpose of the hydro-management system constructed within the El Edén wetland. By augmenting upland soils with nutrients harvested from the wetland, the ancient inhabitants of Makabil could have provided better for their families in an environment otherwise quite inhospitable.

It is worth taking note of how the remains of periphyton are distributed within the local settlement pattern. Snails were only recovered from within the area of the Makabil settlement, not from unoccupied upland areas that would likely have been used as milpas. This suggests that soil augmentation was either only occurring in home gardens or was less concentrated (and therefore statistically harder to find) in milpas. Furthermore, although soil sampling only followed the transect to just inside the boundaries of Cenote Azul, evidence for periphyton was not recovered from within the more elite settlement. It is possible that the use of this wetland product was limited to the residents of smaller peasant communities in closer proximity to its source, or use of periphyton within Cenote Azul was isolated in activity areas that we did not sample.

These results have sparked several new research questions that must be addressed. For example, is there evidence for the use of periphyton at larger, more elite sites or at sites near other wetlands? Was periphyton traded outside of the Yalahau region? How much periphyton can be collected each year and still sustain the resource? Was periphyton used in particular parts of a houselot, such as only in planting beds or in the orchard area described by Heidelberg and Rissolo in this volume? Only the continuation of the scientific process will tell us.

Further testing has already begun at the nearby site of T'isil, drawing evidence from another wetland and a larger site where a variety of potential agricultural areas might illuminate more specifically how periphyton was used. If samples from T'isil indicate that the use of periphyton was not unique to El Edén, regional investigations will be undertaken in order to understand better the significance of this wetland product to the ancient subsistence economy of the northern Maya lowlands.

3

Cenotes, Wetlands, and Hinterland Settlement

Charles W. Houck Jr.

Since the 1960s, the study of prehistoric settlement patterns has formed an integral component of archaeological method in the Maya area. During most of this period, settlement research concentrated on the exploration of large urban centers, providing important information on ancient Maya social and political structure, ritual practices, economy, and the activities and interaction of the ruling elite. Recently, however, a number of projects have begun to combine settlement data from both large sites and their surrounding hinterlands in an effort to gain a more complete view of Classic Maya centers and the polities they ruled. This chapter will examine the methodology and philosophy of hinterland studies, the wide variety of data they can generate—both cultural and non-cultural—and the ways in which the approach has been put into practice at the northern Maya site of Ek Balam. I will also discuss the manner in which the investigation of rural settlement patterns in the northern Maya lowlands has simultaneously helped us to revise our views on the ancient demography of the region and brought to light new data on the distribution of natural resources, especially water-related resources, on which the population depended for survival.

Questions/Problems I: Hinterland Settlement Patterns

The study of the ancient hinterland finds its roots in the strong tradition of settlement pattern research in the Maya area. For over forty years, the investigation of every major Maya center has included an intensive mapping operation, with work at Tikal (Fry 1969; Puleston 1983), Seibal (Tourtellot 1988a), Dzibilchaltún (E. Andrews IV and Andrews V 1980; Kurjack 1974, 1979), and Mayapán (Pollock et al. 1962), among others, providing textbook examples of how analyzing the spatial organization of a population can help us better understand the lifeways of ancient people. In particular, factors such as social and political structure, ethnicity, religion, economy, and subsistence concerns can greatly affect the location, distribution, and organization of human communities. (For an overview of the numerous applications of settlement pattern research see Ashmore 1981; Ashmore and Wilk 1988; Ashmore and Willey 1981; Pyburn 1997.)

For years, archaeological work in the Maya area focused primarily on the civilization's urban capitals, where settlement studies yielded invaluable data on the nature of Maya states and the elites that occupied the upper echelons of society. Site layout, for example, tends to reflect the hierarchical nature of Maya society, with rulers, their families, and other important officials housed in impressive masonry palace complexes near the political and religious heart of the city, while those of lesser rank lived farther out in smaller residential groups (Ashmore 1981; Marcus 1993). The close association of temples, administrative structures, and elite residences at many sites underscores the high level of integration between political and religious power in Maya polities, as does the orientation of public architecture with the cardinal directions and astronomical phenomena (Ashmore 1991; Aveni and Hartung 1986; Bey and Ringle 1989). Beyond this, by documenting large, dense populations at many of these sites, settlement surveys demonstrated conclusively that the Maya constructed cosmopolitan cities, not lightly inhabited ceremonial centers, from which they ruled highly stratified, state-level polities (Marcus 1983, 1993; Webster 1997).

While "center-focused" research led to enormous advances in our knowledge of the ancient Maya—by concentrating mostly on the urban component of society—it has also left our interpretations vulnerable to unintended bias, in that data from major centers alone is insufficient to answer basic questions about the nature of the culture at large, especially the role played by the majority of ordinary citizens living outside of the urban areas. Indeed, since the Maya elites depended on the commoners residing in the surrounding area for their livelihood, the manner in which the inhabitants of so-called "marginal" areas organized themselves bears on issues ranging from polity size and sociopolitical structure to the factors that underlie Maya state formation and decline.

Beginning in the late 1980s, this problem began to be addressed by a new methodological trend that emphasized coupling data from both large sites and more peripheral areas, for the purpose of placing Maya centers within the broader context of the polities they ruled. By incorporating the study of hinterland settlement as a fundamental element, this new approach led to a more integrated view of the culture, where

understanding the relations between the elite and non-elite and between the central and peripheral contribute to a greater understanding
of the whole.

Hinterland Studies and the
Northern Maya Lowlands

As its potential value gains recognition elsewhere in southeastern
Mesoamerica, the study of hinterland settlement has gained a firm
foothold in the archaeology of northern Yucatán. Indeed, efforts to
document regional settlement date at least to the 1970s and the *Atlas
arqueológico del estado de Yucatán* project (Garza Tarazona de González
and Kurjack 1980), which, although technically not an intensive hinterland survey, set the stage for subsequent research. The approach as we
know it today debuted in the 1980s with Dunning's (1992) landmark
study in the Puuc Hills and the work of A. Andrews, Gallareta Negrón,
et al. (1989) with the Cupul survey. By the early 1990s, it had begun
to be adopted more widely by the region's archaeologists. Recent and
ongoing research at Ek Balam (Houck 1998a, 1998b, 1998c; J. Smith
2000), Chichén Itzá (Cobos and Winemiller 2001), Labná (Gallareta
Negrón, personal communication 1995), the Yalahau region (Fedick
1996a, 1996b, 1998b; Fedick and Hovey 1995; Morrison 2000), and the
Chikinchel region (Kepecs 1997, 1999) all incorporate a form of hinterland survey into their research design.

Hinterland Studies: Methodology

The methodology of hinterland survey typically involves intensive reconnaissance of intersite areas, that is areas separated from, or peripheral to, larger centers, with the aim of locating and mapping any
and all signs of ancient settlement. As such, hinterland studies strive
to be both broad in scope and comprehensive in coverage, exploring extensive tracts of land while attempting to document communities ranging in scale from minor centers to the most humble of non-
elite farmsteads. As noted, when coupled with ongoing work within
a large center, hinterland survey ranks as a powerful tool in the study
of Maya polities. A cousin to more conventional regional studies, the
fine-toothed strategy of rural survey can also be applied as a means
of gathering detailed information on the inhabitants of areas without
major sites.

The real-world implementation of hinterland research tends to vary
according to the challenges and opportunities afforded by the local

environment. In areas where thick forest makes walking, much less extensive survey operations, difficult or impossible, settlement remains can be effectively exposed and mapped by cutting transects or a grid of *brechas* (pathways) at regular intervals within a defined area (Fedick and Hovey 1995; Ford 1990, 1991; Morrison 2000; Puleston 1983). Alternately, modern agricultural development in parts of the northern Yucatán has afforded several projects, including my work at Ek Balam (Houck 1994, 1998a, 1998b, 1998c, 2004), the opportunity to survey seasonally burned agricultural fields (*milpas*) and cattle pastures, within which often appear settlement remains that would have been impossible to find in the thorny scrub forest that dominates the region (Glover and Amador 2005; Kepecs 1998, 1999). One should note that, although remote sensing techniques can be helpful in identifying large architecture, the class of remains most frequently found in rural areas—small house mounds—can be difficult to discern when viewed firsthand, much less in aerial or satellite imagery. Simply put, successful hinterland survey requires archaeologists to put their boots on the ground and identify cultural remains in the study area firsthand.

Hinterland Survey: Applications

Regardless of the specific methods employed, the value of hinterland studies lies in the diversity of cultural data they can generate. On a basic level, the dearth of past research in peripheral areas requires most hinterland work to incorporate an aspect of database building, particularly the investigation of important but basic demographic questions. For example, what was the overall population size and density of the rural area in question? When the people formed groups, how large were their communities, and what did they look like? What were the important mitigating ecological or social factors that affected their placement? What was the nature of the elite presence in the periphery?

By assembling a clearer picture of rural demography, we can then begin to evaluate the broader issues of how the Maya organized their states and the complex social, political, and economic forces that shaped them. Specifically, aspects of sociopolitical organization come to the fore through the identification of polity-wide settlement hierarchies, in which formal types of hinterland communities are defined and ranked according to size, settlement density, and other salient characteristics. Since spatial organization tends to reflect cultural organization, the levels within a settlement hierarchy likely indicate important administrative and social divisions within Maya society. Through the judicious analysis of hinterland data, we can, to greater or lesser degrees, gain a more detailed idea as to why these divisions were important, how they related to one another, and what their functions were within the political and economic framework of the state.

In addition to polity research, the high degree of detail typically recorded by hinterland survey makes the data uniquely appropriate for studying the smaller-scale topics of rural household and community organization. Comparing the spatial layout of rural communities and households, both elite and non-elite, with those of their more urban neighbors can shed light on the finer class divisions in Maya society. Moreover, by examining the size, location, and layout of peasant villages and household groups, we can uncover new information regarding the priorities, values, and daily life of the average citizens of the Maya world.

Questions/Problems II: Water and the Ecology of Northern Yucatán

In addition to helping us reevaluate the issues of ancient political and social structure, the techniques involved in hinterland studies have become instrumental in revealing new aspects of the complex relationship the Maya shared with the natural environment. As a result of intensive on-the-ground survey, archaeologists can also recover information on the distribution of natural resources in an area, including highly localized resources that may not appear on maps or in remote imagery. Indeed, factors such as soil depth and quality, vegetation, and average annual rainfall can change significantly (and rapidly) as one moves across a given region. This variation would have influenced both the short-term settlement decisions of the Maya and the higher-order adaptations that allowed them to build and maintain a state-level society. Within the context of northern Yucatán, however, perhaps no single environmental factor influences the size and placement of communities, both ancient and modern, more than the availability of fresh water.

The problem of water in the northern lowlands derives primarily from the region's geology. The Yucatán Peninsula is basically a large platform composed of highly permeable limestone, which permits rainwater to percolate rapidly from the surface to the subsurface aquifer that underlies much of the region (Back 1985; Weidie 1985). As a result, almost no significant sources of surface water—rivers, streams, or lakes—exist on the peninsula north of the Río Hondo. In areas with above-average rainfall, most notably in the Puuc Hills, the collection of rainwater may have been sufficient for survival. Most of the inhabitants of the semi-arid northern plain, however, depended on subterranean water, which they accessed through both artificial wells cut into the bedrock and natural landforms, such as *cenotes* (karstic sinkholes).

Historically, our thinking on water availability in northern Yucatán has centered on cenotes, especially the open-throated, barrel-shaped variety evident at many archaeological sites, such as the Sacred Cenote of Chichén Itzá. On average, this type of columnar cenote measures between 20–50 m in diameter, which allowed convenient access to the water in its bottom and, of benefit to modern researchers, makes them visible from the air (White 1988). The latter is important because all of the region's topographic maps were created using aerial photos, and only physiographic features large enough to be seen from the air actually appear on them. As a result, we have been left with the impression that the distribution of cenotes in northern Yucatán is at best very uneven, with large numbers appearing in some areas and few or none in others. For example, the density of cenotes appears inordinately high in the "Zona de Cenotes," the semicircular region southeast of Mérida where columnar cenotes are extremely common, while the region between Valladolid and Tizimín appears almost cenote-free. Elsewhere, the appearance of columnar cenotes has been linked with the occurrence of subsurface fracture zones, which in the relatively fault-free northern lowlands are fairly random in distribution (Ward et al. 1985). It is possible that the lack of cenotes on existing maps has influenced our past hypotheses regarding regional settlement patterns, in that we would not expect to find large numbers of people living in an area without access to water.

Hinterland Survey and Water Resources

The extensive efforts of hinterland-oriented projects in northern Yucatán have contributed important new information on the distribution and diversity of water-related resources in the region and have given us new insight into many areas of ancient Maya cultural ecology. Data from El Edén (Bell 1998), Chikinchel (Kepecs and Boucher 1996), and Ek Balam (Houck 1998c, 2004; Ringle and Bey 1998), among others, indicate that while columnar cenotes were an important resource and are significantly underreported on maps, they represent only one of several types of sinkhole features (karstic dolines) from which the Maya drew both water and other benefits. As such, I will turn to a brief discussion of the geology of northern Yucatán, the different classes of water-based resources that occur in the region, and the ways in which the distribution and exploitation of these resources may have influenced rural settlement patterns near the site of Ek Balam.

With the exception of the Puuc Hills, most of the northern portion of the Yucatán Peninsula takes the form of a karstic limestone plain whose defining characteristics include flat topography, thin soil (usually less than 1 m), extensive bedrock outcroppings, and the frequent appearance of sinkholes, also known as dolines (Lesser and Weidie 1988; Weidie 1985). In general, the term doline refers to a bowl-shaped depression that occurs in karst limestone environments, usually formed by either the erosion of bedrock by the infiltration of surface water, the collapse of shallow subterranean caverns, or a combination of the two (White 1988). The dolines of Yucatán typically measure less than 100 m in diameter, with depths varying from 2–3 m to below the water table (greater than 19 m). While rounded, bowl-shaped sinks predominate, the region's dolines take a range of forms, including vertical shafts, funnels, and shallow saucers (White 1988).

A Folk Taxonomy of Doline Types in Northern Yucatán. Although very little of the literature on karst geology deals specifically with the dolines of Yucatán, it is clear that at least four formal types occur with frequency: solution sinks, collapse sinks, vertical shafts, and cenotes. Based on the formation processes that create the various doline types, this nomenclature, while useful in describing geologic phenomena, can prove problematic for those of us working on the ground, in that the different forms can have identical surface characteristics and thereby be difficult or impossible to classify. As such, I prefer to use the terminology developed by the Maya themselves, which, although less technical, draws more on observable attributes (specifically morphology and water access) and reflects the function and importance of different sink types within traditional Maya culture. Based on extensive discussions with the Maya farmers of the Ek Balam area, the region's dolines can be divided into three categories: cenotes, *rejolladas* (dry sinks), and *dzadzob* (sinks with swampy bottoms), each of which break down further into several subcategories (fig. 3.1). The terms *cueva* (cave) and *aguada* (depression with potable water) are also used as catchall terms, the former referring to geologic cavities that do not fit neatly into any of the three main types, and the latter referring to any sinkhole with moisture in the bottom (Houck 2004).

Typically the deepest type of doline, cenotes are defined by their penetration of the water table, creating an exposed pool of water that covers the entire floor of the sink. Although some cenote pools appear

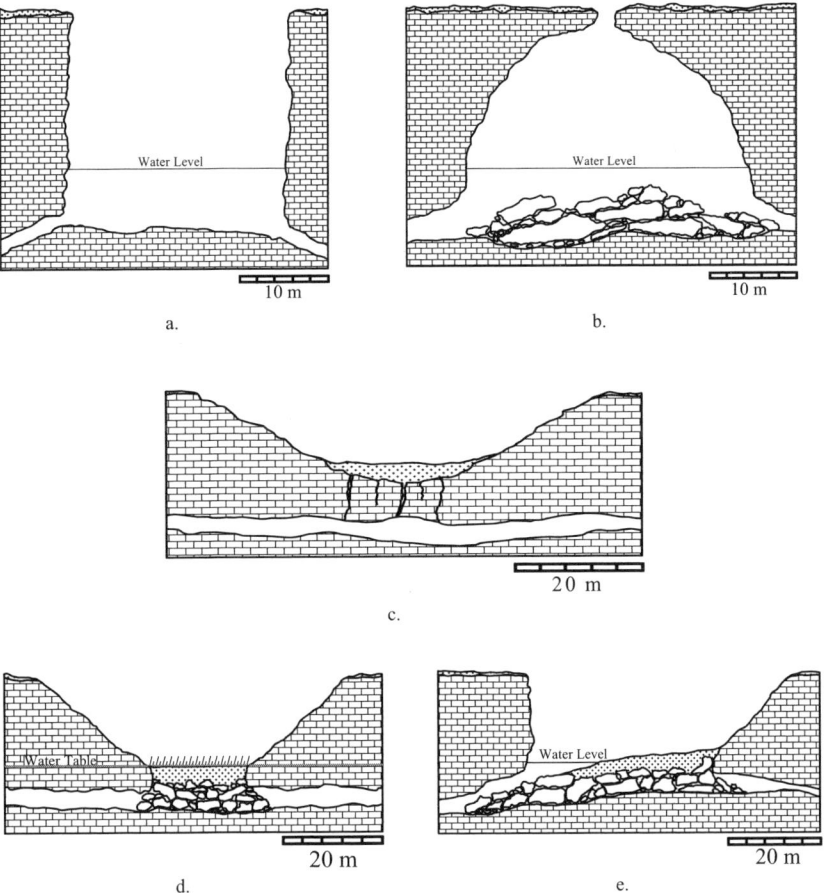

Fig. 3.1 Typical doline forms found in the Ek Balam region: *cenotes*, columnar (a) and *cubierto* (b); *rejolladas* (c); and *dzadzob*, water-only (d) and multi-use (e). (Image created by Charles Houck.)

stagnant, most seem to benefit from sufficient inflow from the surrounding aquifer to maintain a constant freshwater supply. Cenotes occur in one of two forms—columnar or covered (*cubierto*). As discussed earlier, columnar cenotes are the more familiar variety—generally round in shape, with vertical walls that extend from the rim to water surface. Covered cenotes consist of a water pool within a partially collapsed cavern, accessible only through a hole in its roof, such as Cenote Xk'ek'en in D'zitnup, near Valladolid, Yucatán. Although few appear on the maps of the area (they are invisible in aerial photos), covered cenotes outnumber columnar cenotes in the Ek Balam region on the order of two to one and have been frequently used as wells in modern, colonial, and ancient settlements across northern Yucatán (Houck 2004).

Textbook examples of karstic dolines, rejolladas are the defining physical features of the karst plain, with thousands appearing in aerial photographs of the region. Usually round with walls that slope down to a level, soil-covered floor, rejolladas tend to vary more in scale than in form. The various rejolladas documented during the rural survey at Ek Balam range from large, crater-like features greater than 100 m in diameter and 15 m deep to shallow depressions less than 25 m across and less than 5 m deep. Rejolladas were likely important sites to the an-cient Maya for at least two reasons. First, proximity to the water table makes the bottom of a rejollada an ideal location for digging a well. Several examples of stone-lined pre-Columbian wells exist within the limits of Ek Balam as well as in the sustaining area. Second, deep rejolla-das also tend to act as traps for soil and moisture, producing patches of ground useful for cultivating more water-sensitive plants or even small groves of trees (Kepecs and Boucher 1996). Modern Maya farmers com-monly take advantage of these favorable properties, growing tomatoes, pineapples, and bananas.

Technically known as a karst fenster, a *dzadz* is a doline form that touches, but does not significantly penetrate, the water table (National Center for Environmental Assessment 1999). As a result, at least a por-tion of a typical dzadz bottom contains a wet, swampy area, or an ex-posed water pool. Although we encountered a handful of examples near Ek Balam where the entire bottom was inundated, these were the exception, not the rule. Of the three sink types, dzadzob exhibit the widest range of variation in terms of shape and configuration. They can appear with sloped sides (like a deep rejollada), vertical walls (like an underdeveloped cenote), or, most often, a combination of the two. Dzadz sizes vary accordingly with morphology, in that those with sloped sides can exceed the diameter of the largest rejolladas, while those with more sheer walls can approach the range of the smallest ce-notes. Most lie between these extremes.

One should understand that dzadzob represent a relatively rare eco-logical niche for the northern plains, that of an areally restricted wet-land. The value and importance of the dzadz microenvironment to the human inhabitants of the region derives from its versatility as a resource. First, dzadzob provide a dependable water source, either through a standing pool of water or very minor well excavation. In sev-eral dzadzob, we recorded examples of retaining wall construction at the water's edge, presumably to prevent a pool or well from silting in.

As with cenotes, some pools appeared stagnant, while others seem to be refreshed by underground circulation. Second, and perhaps more important than their role as natural wells, the moist, rich soil of the dzadz floor can support a wide variety of plant species unsuited for the drier surface conditions. Even more than with deep rejolladas, Maya farmers could use dzadzob to produce fruit and vegetable crops that could both supplement their maize-beans-squash diet and be used as tribute. Dzadzob may also have been used to cultivate valuable trade crops such as *cacao*, making them a valuable economic resource in the eyes of the state (Houck 2004).

It should be noted that dzadzob can be broken into at least two functional categories, based on the ratio of wetland to dry land contained in their bottoms. Those with a large marshy area or water pool probably provided a good source of water but little in the way of cultivable soil. I will refer to this class as a single-use or water-only dzadz. Dzadzob with a higher percentage of dry surface area, on the other hand, furnish a community with a more versatile resource, combining convenient water access with great agricultural potential. Hence, I label these as multi-use dzadzob.

Current Research:
Hinterland Survey at Ek Balam

Research in and around the site of Ek Balam—located 59 km northeast of Chichén Itzá and 20 km north of the modern city of Valladolid (see fig. I.1)—further illustrates the value and uses of hinterland survey and draws into focus the complex interplay of cultural and economic phenomena operating within the polities of the northern Maya lowlands. Nearly a decade of mapping and excavation in the site center of Ek Balam has revealed massive public architecture and a zone of dense settlement approximately 3.5 km in diameter (Ringle et al. 1991). Ceramic material from surface collections and excavations indicates a long occupation history, extending from Middle Formative times (700–450 BC) to after the arrival of the Spanish in the sixteenth century (Bey et al. 1998; Bond-Freeman et al. 1998). The predominance of Cehpech ceramics, combined with radiocarbon and epigraphic data, place the peak of Ek Balam's development within the Late Classic/Terminal Classic periods, about AD 700–1050 (Bey et al. 1998). Pottery from the Chichén Itzá–associated Sotuta sphere makes up less than 1 percent of the total assemblage from Ek Balam (Bey et al. 1998). This suggests a very low level of interaction between these two major centers, despite an overlap in occupation and their relative proximity to

one another, raising basic questions about how the Terminal Classic/
Early Postclassic Maya polities of Yucatán were organized and related
(Bey and Ringle 1989; Lincoln 1986; Ringle and Bey 1992; see also
Smith et al., this vol.).

The site of Ek Balam has a 9 ha ceremonial precinct surrounded by
a low double wall, containing ten major structures organized around
a rectangular plaza, the largest of which measures 165 m long, 65 m
wide, and 31 m high (Ringle et al. 1991; Ringle and Bey 1995; Vargas de
la Peña and Castillo Borges 1999). Five *sacbeob*, or causeways, extend
from gateways at the cardinal directions, of which three—those to the
north, east, and west—travel nearly 2 km in each direction and are associated with comparably sized elite residential compounds and numerous small structures (Bey and Ringle 1989; Ringle et al. 1991). Residential structures, ranging in complexity from well-defined platforms
supporting multiple dwellings to simple houses on bedrock, densely
pack the landscape for an estimated 12 km^2 (Ringle and Bey 1995).

Simultaneous with the exploration of the site center, members of
the rural settlement survey spent six field seasons investigating the
effect Ek Balam had on the spatial and social organization of its support
population. Specifically, we set out to examine the relative size, density, and distribution of elite and non-elite sites in Ek Balam's sustaining area and how these peripheral communities related to one another
and to the rulers in the center (Houck 1996, 1998a, 2004). The following discussion focuses on this research and its results, with particular
emphasis on the complex interplay of social, political, and economic
forces that shaped the Ek Balam polity during its Late Classic fluores-
cence.

Rural Survey Methodology at Ek Balam

The base methodology of the rural survey involved mapping structures
in cleared ranch pastures and burned milpas currently in use by local
peoples. In a region covered almost entirely by thick, impenetrable
scrub forest (visibility rarely exceeds 5 m), these numerous and widely
distributed fields presented us with expanses of clear ground surface,
exposing even the smallest of archaeological features. To define the
sampling universe better, we concentrated most of our efforts within
a 10 km (east-west) by 8 km (north-south) survey rectangle, situated
3 km due west of Ek Balam. The placement of the survey zone was
largely arbitrary, although the general accessibility of the area, as well
as its association with one of Ek Balam's longer sacbeob, made the western quadrant attractive for investigation (Houck 1998a, 1998c, 2004).

In each ranch or *milpa*, we mapped field boundaries, details of any architectural remains, the location of cultural features, and important topographic details such as rejolladas or water sources. With the aim of recovering sufficient artifacts for relative dating purposes, we performed surface collections (when practicable) on each mapped structure and selected one structure in each site for stratigraphic testing with 1 × 1 m test pits. Finally, each field's location was plotted on a 1/50,000 geologic relief map of the area using coordinates taken with a hand-held global positioning system (GPS).

In addition to our milpa survey data, we targeted two of the more impressive rural centers, Xuilub and Yohdzadz, for more intensive survey and excavation to test the relationship between the peripheral elite population and the rulers of Ek Balam. At Xuilub, we cut a 500 m² grid in the forest north of the previously mapped site center, within which we cleared and mapped all encountered structures. We also excavated a total of twelve 2 × 2 m test pits in three residential structures, including a 3 m deep unit in one of the site's largest platforms, which produced Middle Formative ceramics and the rural survey's only burial (Bey et al. 1998; Bond-Freeman et al. 1998; Houck 1996, 1998a, 2004). In the case of Yohdzadz, we benefited from the site's placement between a large ranch and a small milpa, which we linked using cruciform transects each 100 m wide and 500 m long. This allowed us to document the principal architecture in the site center while exploring the site limits and settlement density within a mapped area of more than 32 ha. As at Xuilub, we selected three structures of varying architectural complexity for test pitting (Houck 1998b, 2004).

Data Recovered

Through three field seasons of preliminary reconnaissance and three more of full-blown survey, we produced a total of sixty-nine maps covering over 411 ha. Of these sixty-nine plans, fifty-one contained at least some evidence of ancient occupation (356 ha). Within the survey zone, we mapped fifty-nine milpas and ranch pastures, covering 336 ha, for a 4.5 percent sample (fig. 3.2). Evidence of occupation was found in forty of the mapped areas in the survey zone. In addition, we excavated fifty-eight test pits and conducted nearly 200 surface collections, resulting in the recovery of over 34,000 potsherds. We also located more than ten additional settlements in the greater Ek Balam area. In terms of actual cultural units, our maps of occupied areas represent portions of at least twenty-one prehistoric communities, which, despite wide variation in architectural sophistication, generally share

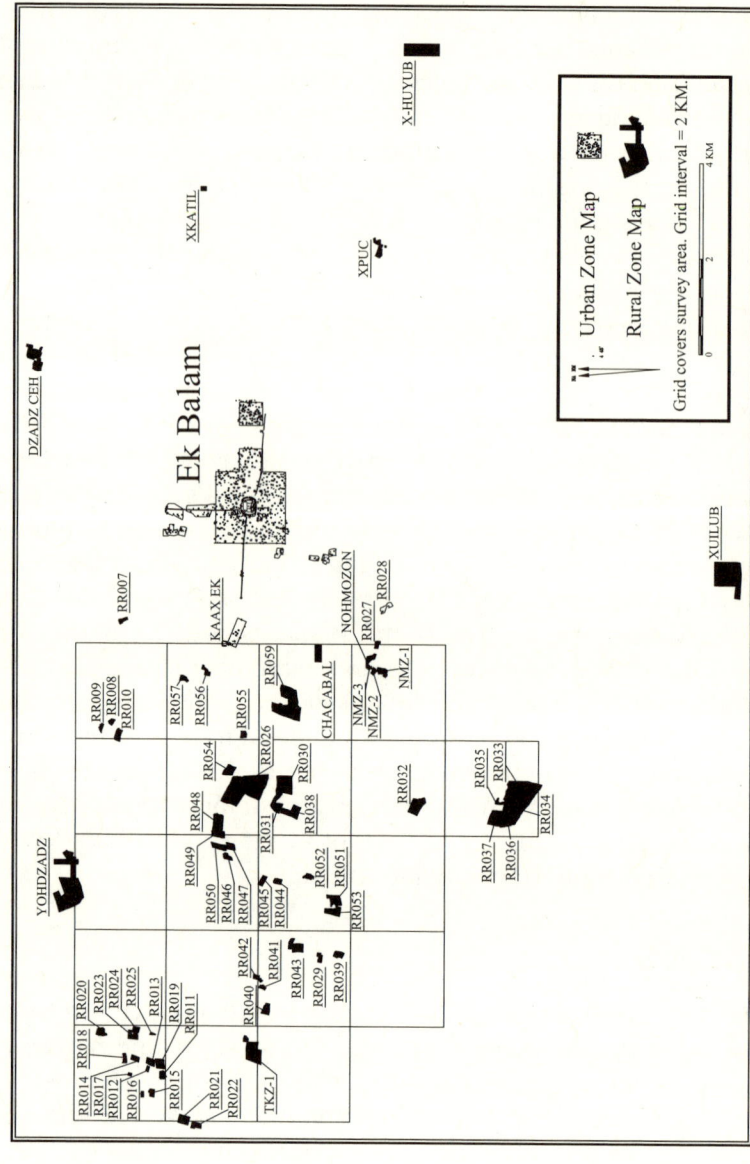

Fig. 3.2 Rural survey of Ek Balam map locations and survey grid. (Image created by Charles Houck.)

the developmental trajectory of Ek Balam and its peak in the Late Classic period (Houck 1998c, 2004; Ringle and Bey 1998).

Given the absence of cenotes on our topographic maps, we were somewhat surprised to find eighty-one discrete water sources, seventy-nine of which fall within a 17 km radius of Ek Balam. In terms of the types and varieties we identified, there were fifty cenotes (thirty-three covered, seventeen columnar), twenty-six dzadzob (five water-only, sixteen multi-use, five of indeterminate variety), and five fabricated wells (fig. 3.3) (Houck 2004).

Settlement Hierarchy

The data amassed during our work to date has led to the formation of a five-tiered settlement hierarchy for the Ek Balam region, with the component levels designed to represent the major social, political, and economic divisions within the Ek Balam polity (table 3.1). In keeping with the rural survey's objective of documenting the frequency and distribution of rural elite and non-elite settlements, sites were first divided into either "elite" or "non-elite" categories based on the presence or absence of public architecture (usually pyramids or especially large platforms). The final rank for elite sites was based on criteria such as number, volume, and height of public buildings; quality of architecture; and presence or absence of a formal arrangement or civic plan—in particular, temple assemblages or plaza groups. Non-elite site rank depended on quantity of platforms and overall community size. It should be noted that, although influenced by them, the levels in the Ek Balam hierarchy do not directly correspond to the site ranks used by larger regional studies like the *Atlas arqueológico del estado de Yucatán* project (Garza Tarazona de González and Kurjack 1980) and the Cupul survey (A. Andrews, Gallareta Negrón, et al. 1989), due both to the smaller scale of our study area and research questions and to the detailed nature of the Ek Balam rural survey data. Concerned with the identification of sites over a wide area, these big surveys depended primarily on aerial photos and informants to locate ruins, skewing the sample toward centers with larger, more noticeable architecture. Focused on a more localized area, the extensive ground reconnaissance employed by the rural survey of Ek Balam allowed the recording of settlement remains as ephemeral as single, often incomplete, house foundations. As such, the site ranks discussed here are specifically tailored to best describe settlement within the Ek Balam region and may not be appropriate in other areas.

Beginning with the top tier, Level I consists of sites in the "capi-

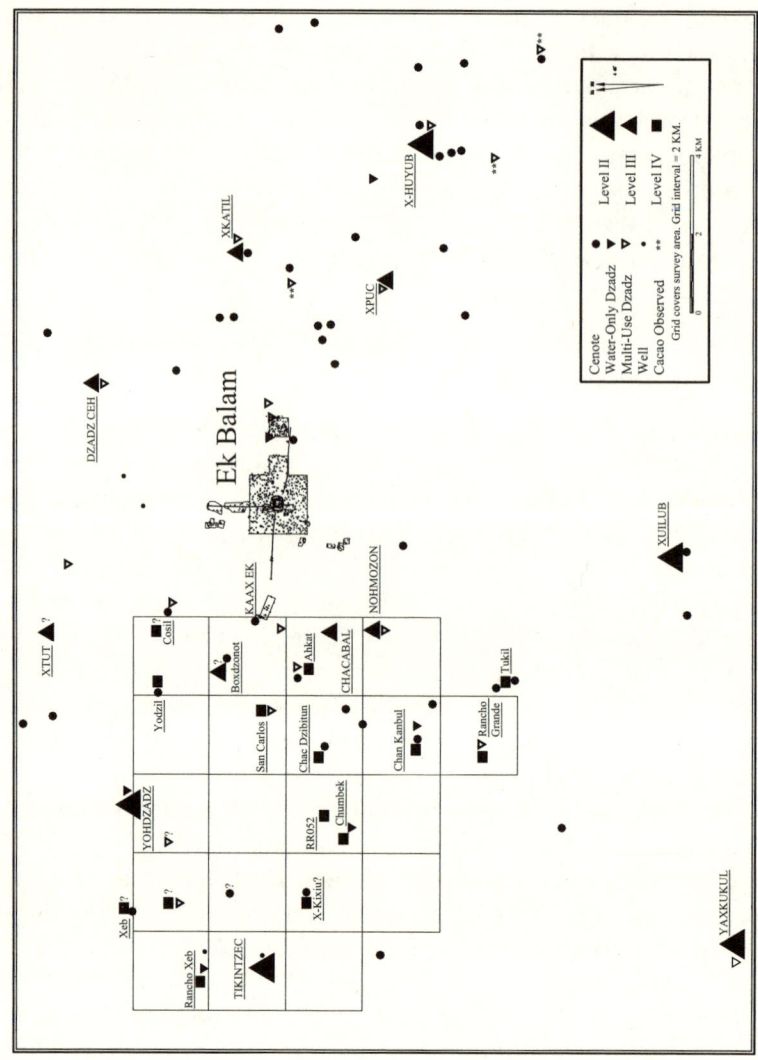

Fig. 3.3 Rural sites and water sources. (Image created by Charles Houck.)

Table 3.1 Rural Communities by Site Rank

Rank	Site Name	Quad	Easting	Northing	Maps
2	Tikintzec	16Q	370200	2310732	TKZ, TKZ-1
2	X-Huyub	16Q	374199	2287898	X-Huyub
2	Xuilub	16Q	386669	2299734	Xuilub
2	Yohdzadz	16Q	374606	2314550	Yohdzadz
3	Chacaba1	16Q	368955	2311568	Chacaba1
3	Dzadz Ceh	16Q	384832	2314925	Dzadz Ceh
3	Nohmozon	16Q	377208	2304715	Nohmozon, NMZ-1, NMZ-2, NMZ-3
3	Xkatil	16Q	388600	2311403	Xkatil
3	Xpuc	16Q	387374	2307626	Xpuc
3/4	Box Dzonot	16Q	377933	2311675	056, 057
4	052	16Q	374124	2309077	052
4	Ahkat	16Q	377599	2309444	059
4	Chacdzibitun	16Q	375495	2309359	038
4	Chan Kanbul	16Q	375640	2306742	032
4	Chumbek	16Q	373580	2308431	051, 053
4	Kosil	16Q	379450	2313150	007
4	Rancho Grande	16Q	375708	2305031	033, 034, 035, 036, 037
4	Rancho Xeb	16Q	369908	2312254	011, 012, 015, 019
4	San Carlos	16Q	376357	2310799	054, 055, 026
4	Tukil	16Q	377394	2304277	Tukil
4	X-Kixiu outlier	16Q	372051	2310221	042
4	Xeb outlier	16Q	370842	2312829	023
4	Xeb outlier	16Q	370851	2312833	024
4	Yodzil	16Q	377200	2312650	008, 009, 010
5	NMZ outlier	16Q	378990	2307585	027
5		16Q	370899	2313523	020
5		16Q	379727	2307482	028
5		16Q	376123	2309658	030
5		16Q	372468	2308403	039
5		16Q	371361	2309973	040
5		16Q	372690	2309425	043
5		16Q	374523	2310805	046
5		16Q	374811	2310800	047

tal" class: numerous huge public buildings, extensive use of cut-stone masonry and vaulting, sculpted and modeled stucco decorations, multiple plaza groups and temple assemblages, and a large support population. No site within the study area exhibits all of these qualities except Ek Balam itself, and as the clear capital of the polity, it alone occupies the top tier of the hierarchy (Houck 2004).

Level II includes those peripheral elite centers with public architecture of more than 7 m in height, which also demonstrate clear formal arrangement around a plaza. Temple assemblage groups similar to those at Ek Balam are often associated with the central architecture. Vaulted structures may appear, and cut-stone architecture is common. Most representative of this level are the sites of Xuilub, Tikintzec, and Yohdzadz, our three most impressive secondary centers. Based on the size and scope of its architecture, I also include X-Huyub as a Level II site, but it should be considered separately from the other sites in the category, as its occupation was limited largely to the Late Formative/ Early Classic periods (Ringle et al. 1989). Despite their architectural sophistication, the Level II sites exhibit lower overall settlement density (an average of .873 structures/ha) than any of the other communities in the Ek Balam area (Houck 2004).

Level III consists of lower-level elite centers or those with public architecture less than 7 m high and whose plaza groups, when present at all, are less strictly defined. Though uniformly lacking vaulted rooms, significant amounts of cut stone in frame braces and platform stairways denote relatively sophisticated architecture and clear elite occupation. These sites also appear to have incorporated somewhat larger support populations, averaging 3.029 strs/ha. Xpuc, Nohmozon, Chacabal, and Xkatil fit well into this category. Dzadz Ceh, with its 6 m high pyramid and well-formed plaza, is a borderline case but appears to have more in common in terms of size and density (2.941 strs/ha) with the Level III centers than those of Level II.

Level IV in the hierarchy consists of nucleated settlements without measurable elite presence—that is, non-elite farming villages. These settlements are composed almost exclusively of amorphous fill-on-bedrock residential platforms, which can support one or more apsidal or rectangular houses. Though widely variable in terms of size and density, the Level IV sites nonetheless average 2.127 strs/ha. The Level IV villages also were by far the most numerous settlement unit encountered, with examples at Chumbek, Chankanbul, Rancho Xeb, Rancho Grande, Rancho Ahkat, Yodzil, Tukil, San Carlos, and Milpa 052 (Houck 2004).

Finally, Level V includes humble farmsteads—ordinarily isolated house mounds or a group of two or three—not closely associated with any observed prehistoric community. Although some appear to have been occupied year-round, many of these structures—the ones in Milpas 027, 028, 030, 040, and 047/050, for example—were likely for seasonal use by farmers commuting from established villages, to judge from impoverished artifact scatters and the lack of features such as metates.

Discussion

Among the first, and least surprising, trends to emerge from our data set was that, in the semi-arid environment of northeastern Yucatán, people tended to live near secure sources of drinking water. Indeed, Ek Balam itself contains at least five water sources: two columnar cenotes, three wells, and at least one dzadz (McCall 1995; Ringle and Bey 1995). Oddly, none occur near the site center, with the cenotes each located over 1.5 km away, one to the east and the other to the west. Outside of Ek Balam, however, we found that settlement not only tends to cluster around water sources, but that the different types of sites represented in Levels II through IV of the settlement hierarchy appear to correlate with specific types of water sources. Beginning with the Level II centers, all four of the sites in this class are associated with water-only sources: Xuilub and X-Huyub each have cenotes, Tikintzec has a well, and Yohdzadz has an eponymous water-only dzadz. Almost all of the Level III elite sites, on the other hand, center on multi-use dzadzob. The only exception, Chacabal, also reportedly has a dzadz nearby, but we were never able to verify it. Finally, of the thirteen Level IV non-elite communities, eleven were arrayed around, or directly associated with, water-only sources—seven cenotes and three water-only dzadzob. In addition, Rancho Ahkat and San Carlos each incorporated a multi-use dzadz (Houck 2004).

This pattern of association between water sources and rural site types seems to reflect, in varying degrees, both the factors underlying the settlement choices of the groups represented in the settlement hierarchy and the function of the communities they built. The least-strong connection between water availability and site location and function occurs in the settlements most closely tied to the capital—the Level II elite centers. Like Ek Balam, none of these larger settlements have cenotes closely associated with their central precincts, giving the impression that availability of water did not comprise a major factor in determining site placement. Rather, I suspect social convention or political necessity wielded more influence over Level II site location.

For example, the respective positions of Xuilub and Tikintzec, approximately 11 km due south and west of Ek Balam, may indicate both the extension to the periphery of the capital's cruciform layout and, more importantly, its underlying cosmological significance. Ek Balam's principal·elites appear to have derived at least some of their ritual authority from spatial associations with the cardinal directions, and it is not unreasonable that their cousins in the hinterland would do the same (Bey and Ringle 1989; Houck 1998c, 2004). As for site function and an explanation of their underwhelming settlement density, our data are unfortunately inconclusive, although several hypotheses invite further investigation.

First, in housing the ruling elites' closest relatives in the hinterland, these sites probably served as administrative centers, nodes of control through which the capital kept an eye on its more peripheral subjects. Second, some may have also served the state's strategic interests, with Xuilub and Tikintzec especially well placed to serve as points of contact with Ek Balam's largest neighbors—Cobá to the south and Chichén Itzá to the west. In both cases, the large, well-constructed public buildings would impress both foreigners and locals with Ek Balam's power, but the sites would not require a large support population to accomplish their official function. Consistent with this view, Ringle and Bey (1998) suggest that these sites may have been "manor communities," country homes or rural strongholds of elites that lived in Ek Balam for part of the year.

In contrast, Level III elite sites appear placed specifically to take advantage of their associated water-related resource, the multi-use dzadz. Indeed, the coupling of elite occupation with multi-use dzadzob suggests that these locations held a certain degree of value within the Ek Balam polity beyond that of simple water. It is not unlikely that the multi-use dzadz environment, with its rare humid soils, formed an important economic resource that merited the close, constant attention of the elite authorities. Moreover, I suspect that Ek Balam exploited these areas for the production of valuable trade crops, including cacao.

While colonial accounts of cacao groves, and even "plantations," in northern Yucatán met with skepticism for many years, recent research concentrating on the cultivation of sinkholes demonstrates the viability of this hypothesis. In particular, the discovery by Arturo Gómez-Pompa of cacao growing in three sinkholes near Valladolid looms large, confirming both the existence of cacao in the northern plains and the suitability of the sinkhole microenvironment for its cultivation (Gómez-Pompa et al. 1990). Moreover, all three of Gómez-Pompa's sinkholes fit securely into the multi-use dzadz classification (Gómez-

Pompa et al. 1990). In the Chikinchel region west of Ek Balam, Kepecs and Boucher (1996) argue convincingly for cacao cultivation in deep, but dry, rejolladas. They also note an association between elite residences and some rejolladas. Finally, during the 1995 season, we identified our own example of wild cacao growing in the bottom of a multi-use dzadz 5.6 km east of central Ek Balam (McCall 1995). While certainly circumstantial at this point, the prospect of significant cacao production in the northern lowlands tickles the imagination and demands further study.

The strong association of Level IV sites with water-only resources indicates that satisfying the basic need for water was clearly of paramount importance in the formation of these non-elite villages. Indeed, most of the non-elite settlements in the Ek Balam hinterland not only located themselves near water sources but also built their communities so that the water source appears at or near the site center. For the ancient farmers who inhabited these sites, this arrangement certainly was the most advantageous, allowing easy access to the largest number of people. In addition to the utilitarian value of cenotes and dzadzob, the presence of small shrines built on or near the dzadz edge at both Chumbek and Rancho Xeb suggests that, like their rulers, "ordinary" Maya also saw these features as sacred spaces and used them for ritual purposes. As such, their water sources occupied the center of both their physical and spiritual worlds.

Conclusions

The example of hinterland research at Ek Balam underscores the value and potential of the approach for the broader study of the ancient Maya world. By combining settlement and environmental data that can only be acquired through the intensive techniques of rural survey, we can develop more complete and accurate models of Maya culture, taking into account the activities and contributions of people from all levels of society. Indeed, in the case of Ek Balam, the hinterland emerges not as a sparsely populated area devoid of activity but as a thriving, culturally dynamic region. Settlements inhabited by people from every stratum of Ek Balam society dot the landscape. Farmers, not unlike the *campesinos* (farmers) of today, lived in small communities near their fields and worked to support their families and meet the tribute obligations imposed by the state. Other settlements saw representatives of the ruling class overseeing the development of valuable agricultural lands, which potentially produced equally valuable crops. Though somewhat distant from the capital, close relatives of the Ek

Balam power elite lived in sumptuous surroundings, maintaining by their presence the authority of the center in the surrounding area.

While our work at Ek Balam has revealed much about the polity at large, and has given us insight into the interaction of its central and rural populations, broader questions remain that we can only address through continued hinterland research, both in the northern lowlands and throughout the Maya area. For example, how do the patterns found at Ek Balam compare to those at other sites? Are the Ek Balam settlement types, and the social organization they reflect, unique to the one site or common at other major centers as well? If similarities exist, what are the social, economic, or environmental factors that cause them? If, on the other hand, Maya polities exhibit a greater range of organizational variation than currently thought, how do we explain it, and how does it affect the way we view Maya states in general? In terms of ecological data, we need to continue to document the type, frequency, and distribution of water-related resources available to the Maya of Yucatán, given their practical, religious, and economic value, and their influence on the region's settlement. Beyond water, however, we need to begin to see the Maya landscape with new eyes, in order to identify other natural resources that the ancient inhabitants considered important but that may have gone unnoticed by earlier investigators. Certainly the answers to these and other questions remain unclear at the moment, but as hinterland research progresses, the data recovered will enable us to create ever more accurate models of ancient Maya political structure and interaction.

4

The Archaeology of Urban Houselots at Chunchucmíl, Yucatán

Scott R. Hutson, Alíne Magnoní, Daniel E. Mazeau, and Travís W. Stanton

The archaeological site of Chunchucmil, located on the northwest corner of the Yucatán Peninsula (see fig. I.1), lacks the massive pyramids, monumental plazas, and ornate stone carvings that typify sites like Uxmal and Chichén Itzá and therefore falls into the second rank of sites classified by the *Atlas arqueológico del estado de Yucatán* (Garza Tarazona de González and Kurjack 1980). However, initial work at the site (Vlcek et al. 1978) recognized Chunchucmil's substantial residential population, and the most recent research has documented settlement density unprecedented for a Classic period lowland Maya site. This research, conducted as part of the Pakbeh Regional Economy Program (PREP) directed by Bruce Dahlin and Traci Ardren, has identified more than 4,000 structures in an intensively mapped area of 6 km². Furthermore, remote sensing indicates that the area mapped thus far represents only a fraction of the total site. The PREP, begun in 1993, has focused on the question of how such a large population could have sustained itself in an area of notoriously marginal agricultural productivity. In this chapter, we begin by reviewing some of the insights gained by the PREP in this regard and then move to a more detailed discussion of the residential compounds that comprise the vast majority of Chunchucmil's dense settlement. We compare these residential compounds to the model of corporate groups proposed by Hayden and Cannon (1982) and use mapping and excavation data to discuss the range of variation among these compounds.

Chronology, Demography, Economic Diversification, and Site Configuration

The earliest evidence of occupation at Chunchucmil dates to the Middle Formative period, but Chunchucmil did not attain its massive size until the Early Classic period. Based on ceramic cross-ties to other local sites (Jiménez Alvarez 2002; Varela Torrecilla 1998), Chunchucmil's population was at its highest during the late part of the Early Classic period and the early part of the Late Classic period (Mansell and Bond-Freeman 2002). Populations dropped in the latter part of the Late Classic period. In the Terminal Classic period, the site was occupied by a small population that resided on broad platforms as opposed to the patio groups common during Chunchucmil's apogee (Magnoni, Hutson, and Stanton 2002).

Chunchucmil does not have clear site boundaries (Ardren et al. 2003; Hutson et al. 2000), therefore making it very difficult to assess the number of people that occupied the site during its pinnacle. Along the western and southwestern edges of our map, structure density drops from approximately 600 residential structures per square kilometer to less than 300, thus allowing us to delimit a boundary between "urban" Chunchucmil and its outskirts (Hutson et al. 2001). We have not yet pinpointed such a boundary to the north, east, or southeast, but aerial photos and Landsat 7 imagery, analyzed by project member Dave Hixson, suggest that Chunchucmil may have covered 25 km². Though we are not comfortable providing specific population estimates that include areas that have not been intensively mapped, the site would have housed more than 30,000 people by any reckoning (Ardren et al. 2003). These data make Chunchucmil the most densely occupied Classic period Maya site, with the exception of the six hectares of Copán's urban core. Of greater interest than cross-site comparison, however, is the consequence of such a large city in such a marginal agricultural environment: the regional population could not have been agriculturally self-sufficient (Beach 1998). It must have depended on alternative resources.

Located at the eastern edge of a seasonally inundated savannah and 27 km from the coast, the residents of Chunchucmil could have procured various foods not available further inland in the region, such as fish, shellfish, birds, jaguar, manatee, monkeys, and crocodiles. During the middle of the Classic period, Chunchucmil was the only large site close to the Celestún salinas, the second-largest salt works in all of Mesoamerica. Chunchucmil's proximity to the Gulf Coast also enabled it to take advantage of one of Mesoamerica's most vigorous maritime trade routes (A. Andrews 1990b). The coastal site of Punta Canbalam may have been Chunchucmil's seaport (Dahlin et al. 1998). We envision the residents of Chunchucmil both as middlemen in the trade of preciosities between highland Mesoamerica and contemporaneous inland sites such as Oxkintok (27 km to the east of Chunchucmil; Rivera Dorado 1987, 1989, 1990, 1992) and as active harvesters of salt and other coastal/savannah commodities that were both consumed locally and traded inland for additional food products (Dahlin and Ardren 2002; Stanton et al. 2000).

Oddly, Chunchucmil contains little evidence of centralized political authority and its accompanying constellation of dynastic succession, broad-scale public rituals, and massive corvée labor projects. In its place, Chunchucmil has fifteen medium-sized temple compounds dispersed over many square kilometers with all but one connected to each other through a system of *sacbeob* (roads)(fig. 4.1; Dahlin et al. 1999; Hutson et al. 2000; Magnoni et al. 2000; cf. Cobos and Winemiller 2001). These compounds have an identical architectural layout, consisting of a patio with a tall pyramid (between 8 and 17 m, averaging 11 m high) on one side, low range structures (approximately 2 m high) on the other three sides, and a performance platform in the center. Some of these groups, called quadrangles (see fig. 4.1; G. Andrews 1985), are connected to additional patios with residential attributes (Blackmore and Ardren 2001). The central patios of the quadrangles measure 3,000 m² on average, and thus no patio could have hosted much more than about a tenth of the site's population. However, the quadrangle with the tallest temple also contains the site's only ballcourt, suggesting the existence of rank among the quadrangles. Though we documented open spaces in the site core, no ceremonial architecture frames these spaces, which may have instead served as market places.

Residential Chunchucmil and Corporate Groups

Chunchucmil's residential zones are remarkable not just for their density of settlement but also for the prevalence of circulatory and boundary-marking features. As in most Maya sites, domestic architecture at Chunchucmil conforms to the patio group model (Ashmore 1981; Willey and Bullard 1965) in which two or more structures face onto a common patio. Yet, at Chunchucmil, low, winding stone alignments referred to as *albarradas* encircle the patio groups (fig. 4.2), creating bounded houselots much like those found in historic and modern Yucatecan pueblos (Hanks 1990:95, 313; Ortega et al. 1993; Restall 1997:99). Reflecting the extraordinarily dense settlement patterns of urban Chunchucmil, there is very little unclaimed or unbounded space within the residential zones. Bounded patio groups often abut each other, like the cells of a honeycomb, sharing a common albarrada wall (see fig. 4.2). Narrow alleyways (2–3 m wide), marked by albarrada

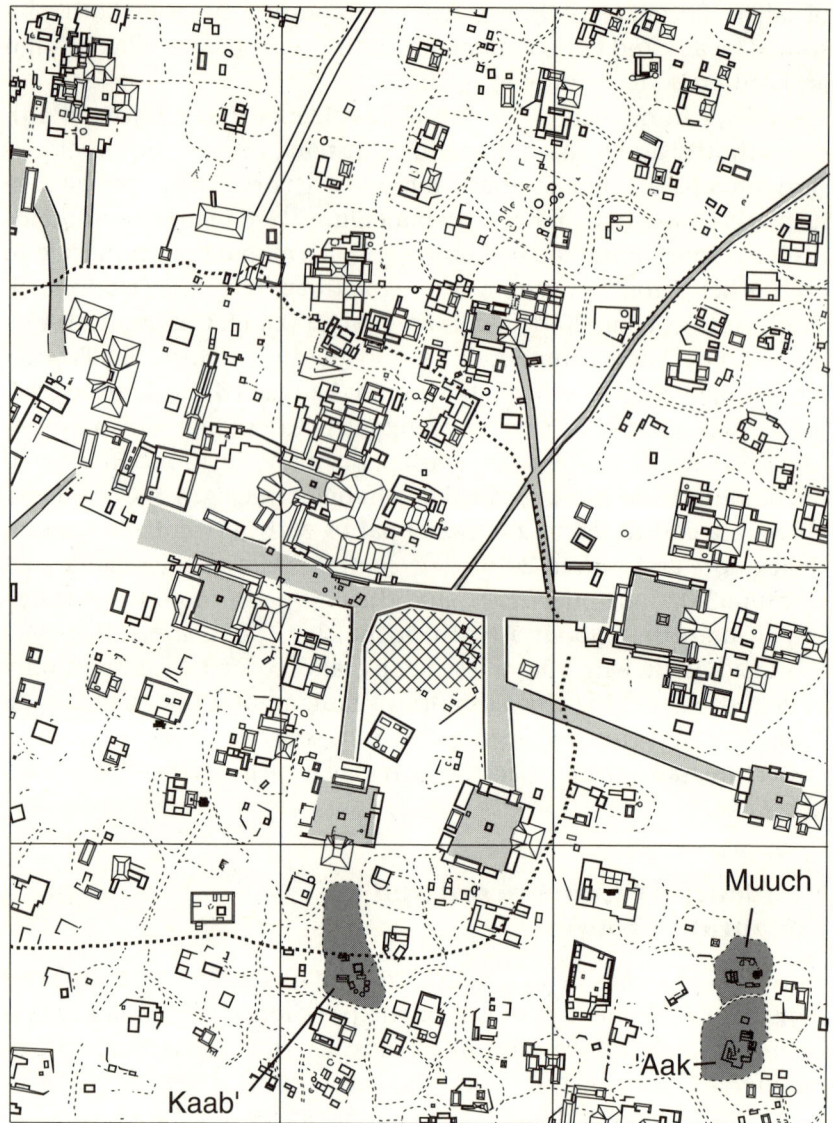

Fig. 4.1 Central Chunchucmil, with quadrangle groups and sacbeob in light grey. The 'Aak, Muuch, and Kaab' groups are highlighted in darker grey. A potential marketplace is marked with cross-hatches. The grid is oriented to north, and each grid square measures 250 m by 250 m. (Map created by Scott Hutson.)

walls on each side, often snake in between two houselot boundaries, connecting cramped neighborhoods with causeways, temples, and open areas in the site center and beyond. Though residential albarradas are common at Postclassic sites like Mayapán (Bullard 1952), Tulum (Benavides Castillo 1981; Vargas et al. 1985), Xcaret (E. Andrews IV and Andrews 1975), Playa del Carmen (Silva Rhoads and Hernández 1991), Xamanha (Goñi Motilla 1998), and Cozumel (Freidel and Sabloff 1984; Sierra Sosa 1994), they are not as common during the Classic period. Furthermore, residential albarradas at Classic period sites such as Cobá (Fletcher 1983; Manzanilla 1987) do not form alleyways for circulation and are often not shared between houselots. Most of the bounded residential groups at Chunchucmil contain multiple residences, as well as a shrine on the east side of the patio that created and embodied for each group a corporate sense of identity (Hutson et al. 2004; see also Becker 1991).

Brian Hayden and Aubrey Cannon (1982:134–35) proposed that corporate groups "come into being as a result of strong economic or environmental pressures, and which, as a result, exhibit a recognizable degree of residential coherency among two or more nuclear families" (see also Manzanilla 1996). The residential compounds at Chunchucmil fit Hayden and Cannon's idea of corporate groups very well: located in an area of strong economic and environmental pressure due to poor farmland, the Chunchucmil houselots contain multiple-family compounds whose boundary walls and shared shrines give them a high degree of residential coherency. The houselots at Chunchucmil correspond to Hayden and Cannon's second type of residential corporate group, in which each family occupies its own structure and all structures are placed near each other in a patterned fashion. In the remainder of this chapter, we focus on documenting and explaining variation between corporate groups at the scale of the houselot.

Houselot Variation: Goals and Methods

The goal of our analysis of corporate groups at Chunchucmil is to use the material residues of houselots to document variation between groups and then to test hypotheses that may help explain this variation. Of the many axes of variation between groups, we explore five in this chapter: (1) the size of the houselot; (2) the distance from the houselot to the site center; (3) the cost of constructing the buildings within each houselot; (4) social coherence within houselots; and (5) economic specialization.

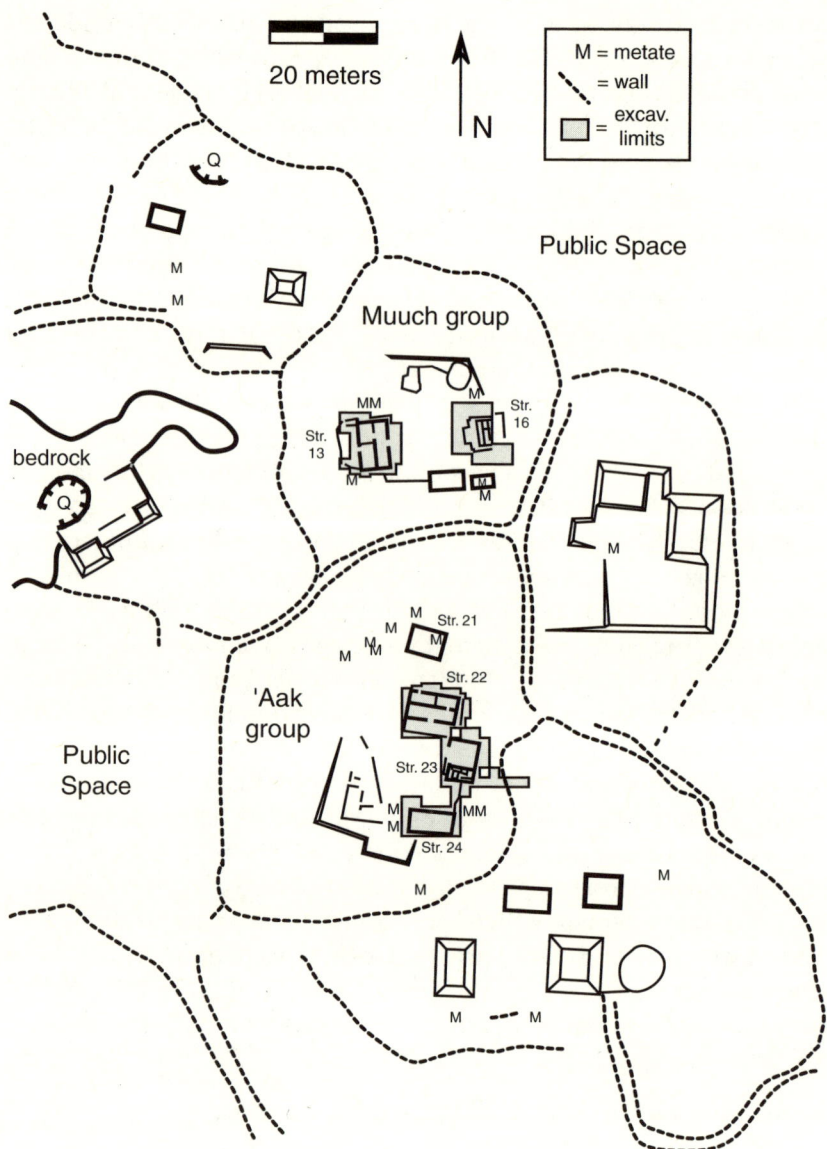

Fig. 4.2 The neighborhood of the 'Aak and Muuch groups, showing the location of excavations as well as an example of how albarrada walls serve as houselot boundaries and form public alleyways. (Map created by Scott Hutson.)

We use two methods to document variation along these axes: mapping and excavation. We have thus far mapped approximately 850 patio groups at the site. Of these 850, less than half are completely enclosed by albarrada walls. We restricted our analysis to completely enclosed patio groups because calculations of houselot size require the houselot to be fully bounded. Unenclosed patio groups share the same layout, scale, and features as enclosed patio groups, thus permitting us to assume that the two types of groups represent the same range of activities. For this analysis, we used a sample of seventy-five houselots found in a 3.1 km² swath (1.2 km north/south by 2.6 km east/west) beginning in the site center and extending to the western edge of the site. Sampling from such a swath ensured that groups from both the center and the edges of the site would be represented. We have conducted test pits or broader excavations in twenty-nine of the groups in the current sample of seventy-five. Analysis of ceramics from these excavations shows that occupation in each of these groups began during the late part of the Early Classic period and, in a minority of the groups, continued into the Late Classic period. We therefore feel justified in assuming that the bounded patio groups at Chunchucmil were contemporaneous in the late Early Classic period. This contemporaneity is further suggested by the way in which these houselots share albarrada walls and the way in which alleys connect multiple houselots: they all form part of an integrated, though decentralized, plan.

In examining the maps of each of the seventy-five houselots, two of the five axes of variation—size of houselot and distance from site center—are relatively easy to quantify. Assessing cost of building construction, social coherence, and economic specialization, however, is more difficult. To assess cost of construction, we use the volume of mounded architecture. The cost of construction is itself a proxy for the prosperity of the social group, since building large architecture requires control over large amounts of labor and resources (Abrams 1994). The difficulty is that both of these leaps, first from mound volume to cost of construction and second from cost of construction to prosperity of corporate group, are indirect.

The first leap is indirect because mound volume is not the only measure of construction cost. Mound volume does not take into account the cost of perishable superstructures. The second leap is indirect because cost of construction is not the only measure of corporate group prosperity. Access to precious goods also contributes to prosperity, and there are cases at Chunchucmil (Ardren and Hutson 2002) and elsewhere (Hendon 1992:36) in which architecturally modest residential complexes have unexpectedly rich burial goods.

The leap from cost of construction to group prosperity is also problematic because construction cost can be spread over many generations. For example, if a large mound were built incrementally, with slight enlargements every generation, the builders would have only needed to control small amounts of labor and resources at any one time. On the other hand, if the same mound were built all at once, this would indeed have required greater control of labor and resources. This equivocality foregrounds the fact that our measurements for each group represent only one snapshot, namely the final one, in what was likely a series of dynamic, intergenerational cycles in the reproduction of households, as documented in Haviland's "musical hammocks" model for Tikal (1988; see also Evans 1993; Tourtellot 1988b). Unfortunately, the quantity of excavation required to assuage all of these concerns for each of the seventy-five groups is not feasible.

The problems in making the leaps from mound volume to construction cost and thence to group prosperity remind us of Christopher Chippindale's (2000) point that there is often a wide gulf between the data that we grasp, or capture, and the archaeological affair we wish to study. Chippindale argues that we should replace the word "data" with "capta" because data gives the mistaken impression that the information we study has been given as opposed to captured (cf. Hayden and Sansonnet-Hayden 2001). In this particular essay, the information we use is certainly more "captured" than given, and to some degree, as Chippindale would argue, what we have captured is a different beast then the one we desire.

By social coherence, the fourth axis, we refer to the degree to which individual families were integrated and unified with the other families of the group. The organization of domestic space serves as a measure of group coherence. Richard Wilk (1983, 1984) proposed that households forming a cooperative union build their houses close together and in conformity to a coherent plan. For Chunchucmil, this might mean that architectural groups whose buildings share a common platform or are aligned with each other at right angles are more coherent (see also Lohse and Hudler 1997). Wilk and Netting (1984) have suggested that the more coherent groups are also the more prosperous. We will test this correlation below.

The fifth axis of variation, economic specialization, is perhaps most difficult to assess based on mapping data alone. Though accurate documentation of specialization requires extensive excavation, mapping data suggest two economic specializations: harvesting of *sascab*, a nutritional additive to corn that is also used for making plaster (Littman 1958); and specialized grinding using grinding stones known as *me-*

tates. Sascaberas (cave-like pockets within bedrock that provide sascab) indicate the presence of sascab harvesting, whereas large concentrations of metates (up to fifteen per patio group) may indicate specialized grinding.

Relative to mapping, horizontal excavations provide direct data on economic specialization and, as mentioned above, can serve as a corrective to errors in leaping from volume of architecture to corporate group strength and prosperity. In this chapter, we consider data from three horizontally excavated houselots: the 'Aak group, the Muuch group, and the Kaab' group. The 'Aak group and the Muuch group are next-door neighbors located about 600 m southwest of the site center (see fig. 4.1). Both began with one residence (presumably housing one nuclear family) but expanded over subsequent generations into multiple residence groups with three or four nuclear families. Excavations in the 'Aak group focused on open spaces and three of the five buildings and cleared 120 m^3, spread over 364 m^2. Excavations in the Muuch group focused on open spaces and two of the five buildings, and cleared 92 m^3, spread over 234.5 m^2. The Kaab' group, located approximately 400 m south of the site center (see fig. 4.1), had more buildings than the other two groups and a longer occupational history, distributed between two separate patios. Excavations cleared 165 m^3 spread over 353 m^2, including two completely cleared structures and six partially cleared structures.

Houselot Variation: Results

The first axis, houselot size, defined as the area enclosed by the albarrada walls minus the surface area of that group's architecture, exhibits wide variation. Houselot size ranges from 753 m^2 to 7,500 m^2, with a median of 2,528 m^2. The Chunchucmil houselots are rather large compared to those of Cobá, for example, where the largest houselot measures 2,286 m^2 (Fletcher 1983). A histogram of the Chunchucmil houselot sizes shows three distinct modes: one at 2,000 m^2, the second at 3,500 m^2, and the third at 6,500 m^2. This variation in size may respond to a number of factors. An elementary hypothesis is that houselot size is directly related to distance from the center of the site, the second axis of variation. Patio groups closer to the edge of the site, where settlement is less dense and land is more abundant, might be larger. We discard this hypothesis because our capta/data show no correlation between these variables (Pearson's r = 0.163, p = 0.161).

Though distance from the site center shows no relationship to houselot size, distance from the site center may relate to our third axis

of variation: cost of architecture. At Dzibilchaltún and Sayil, Kurjack
(1974) and Tourtellot and Sabloff (1989), respectively, found that residences that required more effort to build were located closer to the site
center. Assuming that the elaborate temples and palaces contained in
site centers are seats of authority and sacred knowledge and that impressive residences house people of higher status, the clustering of impressive residences around site centers can be explained by suggesting that residing close to site centers causes one to be associated with
the authority and sacredness therein. Following this reasoning, a residence's distance from the site center indexes the status of the group
living there. At Tikal, however, no relationship holds between a residence's distance from the center and its construction costs (J. Arnold
and Ford 1980). The data from Chunchucmil show a slight but not
significant negative relation between the cost of architecture and the
distance from the site center (Pearson's r = –0.111, p = 0.345). In other
words, houselots with costly architecture are just as likely to be on the
edge of the site as at the center.

Variation in houselot size could be understood better with reference
to gardening. Intensively cultivated houselot gardens would have been
a necessary and crucial complement to subsistence at Chunchucmil,
and it appears that the albarrada walls are, in part, a response to the
corporate group's need to stake out valuable infield garden space. In
three of the four ancient houselots at Chunchucmil that were subjected to systematic, intensive phosphate analysis, we encountered
areas of high phosphate readings in non-mounded areas with little
durable inorganic refuse and relatively deep soils. These lines of evidence suggest artificially fertilized gardens (Hayden and Cannon 1982;
cf. R. Alexander 1999). Though the fourth houselot did not show evidence of phosphate enrichment in its potential garden areas, this lack
of enrichment does not rule out the presence of a garden or mini-
orchard: it merely shows that the area was not intensively fertilized.

If the size of the houselot is a measure of the amount of potential
garden space, a valuable asset in this soil-scarce region, then size of
houselot might be equated with corporate group prosperity, and we
might expect a correlation between houselot size and the third axis of
variation, which we take to indicate corporate group prosperity: volume of architecture. But in those houselots that have not undergone
phosphate testing, is the "capta" of houselot size truly data for assessing garden size? Non-mounded space in houselots may also be for craft
production (P. Arnold 1990) and feasting (Robin 2002:260), among
other things. However, when gardening takes place, the garden area

often absorbs more than 60 percent of the total houselot space (Killion 1992). If feasting space, as opposed to garden space, is what makes a houselot large, then a large houselot would still indicate corporate group prosperity since the more prosperous groups can be expected to sponsor regular feasts (Evans 1993:180; Hirth 1993:123). Finally, before checking for a correlation between houselot size and volume of architecture, we must note that these two variables are not entirely independent: structures with more volume will take up more space if that extra volume is distributed horizontally as opposed to vertically. However, such interdependence would be very weak because, on average, architecture occupies less than one fifth of the total space within the albarrada walls. Turning to the data, the correlation between the two variables is positive and statistically significant (Pearson's $r = 0.378$, $p = 0.001$). This result recalls research at Sayil (Smyth et al. 1995; Tourtellot and Sabloff 1994) where the residences that required the most construction labor had the most land and the most consistently fertilized land around them.

Turning to the fourth axis of variation, social coherence, we tested the idea, noted by Wilk and Netting (1984), that the more coherent groups are also the more prosperous. To test this hypothesis, we split the seventy-five groups into those with a full central platform upon which most structures rest ($n = 24$) and those without any central platform ($n = 16$; of the remaining thirty-five groups in the sample of seventy-five, thirty-two have partial platforms, and three were not easily coded due to modern disturbance), and compared these groups using two potential indicators of prosperity: architectural volume and houselot size. T-tests revealed no significant difference between the two groups for either variable (architectural volume: $t = 0.794$, $p = 0.432$; houselot size: $t = 1.873$, $p = 0.069$), though it is worth mentioning that the groups without platforms had a mean houselot size of 2,300 m², whereas those groups with platforms had a mean houselot size of 3,300 m². We also split the seventy-five groups into those with buildings aligned with each other at right angles ($n = 52$) and those with unaligned buildings ($n = 18$; the remaining five patio groups in the sample of seventy-five were too disturbed to assess alignment). We assume here that in a group whose residences are not aligned with each other, the families are not as closely tied (Wilk 1983, 1984). T-tests revealed significant differences between these two samples for both houselot size ($t = 2.697$, $p = 0.009$) and architectural volume ($t = 2.735$, $p = 0.008$). In other words, patio groups whose buildings were not aligned to each other had smaller houselots (2,200 m² vs. 3,400 m²) and much less costly architecture (173 m³ of fill compared to 587 m³ of fill). In sum, these capta/data indicate that alignment of architecture at Chunchucmil is a good indicator of corporate group prosperity.

We now move to the fifth axis of variation: economic specialization. The two economic specializations detectable by surface mapping include sascab harvesting and specialized grinding. Of the seventy-five groups, eighteen have sascaberas within their albarrada walls, while fifty-seven do not. A t-test revealed no significant difference in mound volume between those houselots with sascaberas and those without ($t = 0.461$, $p = 0.287$). Having a sascabera also has no significant effect on houselot size ($t = 0.206$, $p = 0.838$, df = 73). These results indicate that corporate groups with sascaberas as an economic resource are not more prosperous than groups without this resource.

As for specialized grinding, metates may have been used to prepare *achiote*, to make beer or wine, or to grind corn, salt, pigments, or *cacao* (Watanabe 2000). Trace element analysis at Chunchucmil suggests that some metates were used for the preparation of pigments (Magnoni, Hutson, and Beach 2002). Large numbers of metates might be an indicator of a productive specialization, though they could instead reflect the length of time a patio group was occupied (the longer the occupation, the more metates) or the amount of people living in the houselot. The only variable that is significantly correlated with metates happens to be surface area of architecture (Pearson's $r = 0.283$, $p = 0.027$). If surface area of architecture is understood as an indicator of the amount of people living in the houselot (Kramer 1979; Narroll 1962), then the positive correlation with metates should perhaps be expected: with more people living in a group, more corn must be ground and thus more metates consumed. Data from the three horizontally excavated houselots suggest that the quantity of metates also relates to how long the houselot was occupied. The Kaab' group, which had the longest occupation, had eighteen metates, of which between thirteen and fifteen were in use during the middle of the Classic period. The Muuch group, which had the shortest occupation, had five metates. The 'Aak group, which was occupied for longer than Muuch but shorter than Kaab', had eleven metates. In summary, mapping and excavations suggest that the quantity of metates found in Early Classic residential groups relates not to specialized production but to the amount of people living in the groups and the length of time they were occupied.

Looking closely at these three horizontally excavated groups enables a more thorough examination of the issue of economic activities. Thus far, the strongest evidence for specialized economic activity comes from the 'Aak group, which consumed an extraordinary number of obsidian prismatic blades. We have recovered 670 obsidian artifacts from 'Aak thus far, comprising nearly half of all the obsidian found at the site (Mazeau 2001). Debitage from the production of obsidian prismatic blades in the 'Aak group as well as the other two groups was rare, suggesting that the specialization was not in manufacturing blades but in a task that used blades. There are significantly more blade fragments in 'Aak than in the other groups. At 'Aak, we found 5.6 blades per m^3 of excavations, compared to 2.36 blades per m^3 for Kaab', and 0.38 blades per m^3 for Muuch. Such summary figures might be misleading, however, because they do not account for different types of excavated contexts. For example, at the Muuch group, almost half of the excavations concentrated on the group's shrine, whereas in the Kaab' group, only a quarter of the excavations focused on the group's shrine. After controlling for context (residential versus ritual structures, platform fill versus architectural fill, midden versus plaster floors, etc.), we found the same result: 'Aak yielded more obsidian blade fragments per context, across the board. In all three groups, obsidian was most abundant in off-mound trash deposits. In the 'Aak group, a midden covering 2.5 m^3 (2 percent of the 120 m^3 excavated in 'Aak) accounted for 25 percent of all the obsidian recovered in the group.

The quantity of obsidian at 'Aak is also high relative to the quantity of other basic artifacts, such as household pottery (cf. Sidrys 1977: 100). If there were no specialization using obsidian, the ratio of obsidian to ceramics in the 'Aak group would be equal to the ratio of obsidian to ceramics in the other two groups. On the other hand, if the obsidian served a special purpose in the 'Aak group, then the ratio of obsidian to ceramics in that group would be higher than the ratio of obsidian to ceramics in a group that did not need extra obsidian for specialized production. Thus, we created a ratio of obsidian density (blades per m^3) to ceramic density (kg of sherds per m^3) for each of the three groups. We found that the ratio for 'Aak was about 5 blades per kg of sherds, compared to 0.98 blades per kg of sherds in Kaab', and 0.8 blades per kg of sherds in Muuch. This discrepancy suggests specialized production in the 'Aak group.

On the other hand, Rathje (1983:29) suggests that a high ratio of obsidian to ceramics implies a higher level of material well-being. In other words, it is possible that the occupants of the 'Aak houselot consumed more obsidian because they were wealthier and could acquire it more easily than other houselots. Though the construction of the 'Aak group architecture did not require as much control of labor or materials as in the Muuch or Kaab' groups, two burials within the 'Aak group temple contained wealthy offerings, including eighty-five greenstone and shell ornaments (Ardren and Hutson 2002).

If the 'Aak group's privileged access to exotic materials explains their glut of obsidian, and the obsidian had no special use, then the 'Aak group members could have afforded to throw away blades before they were completely useless, whereas people with less access to obsidian would have had to make each of their blades last longer. If this were the case, we would expect much heavier use-wear on the blade fragments from those groups with less access to obsidian. Macroscopic analysis of wear patterns shows that the blade fragments from the 'Aak group were not as heavily used as those from the other two groups. For example, in the 'Aak group only 8 percent of the blades showed heavy use wear, compared to about 14 percent in the other two groups. Also, in the 'Aak group 31 percent of the blade fragments had no macroscopic traces of wear, compared to 25 percent in the other two groups. Though these differences are notable, they do not strike us as being significant enough to argue that wealth differences alone account for the high quantities of obsidian at 'Aak. Though Prudence Rice (1987) has suggested that obsidian was a wealth good in the Maya lowlands during the Early Classic period, this presumes that chert was abundant as a substitute. At Chunchucmil, however, we recovered very little chert, suggesting that obsidian was not a privileged tool material, but rather a necessary tool material (Mazeau and Forde 2003).

The blade fragments from the 'Aak group are also wider and thicker than the blade fragments from the other two groups, indicating earlier access to obsidian cores. The difference in width between the blades of the 'Aak and Kaab' groups was statistically significant at the $p < 0.001$ level ($t = 7.83$). The differences in thickness between the blades of the 'Aak group and the blades of the Kaab' and Muuch groups were also statistically significant (for Kaab', $t = 6.55$, $p < 0.001$; for Muuch, $t = 2.51$, $p < 0.02$). We have no data on blade length because no complete blades were recovered.

To summarize, we have demonstrated that differences in extent and context of excavation, chronology of occupation, intensity of occupation, and wealth do not completely explain the different rates of consumption of obsidian at Chunchucmil. We therefore conclude that the members of the 'Aak group consumed more obsidian partly because they were involved in a specialized activity using prismatic blades. The nature of that activity has yet to be elucidated. Though residue analyses for starch grains and phytoliths have yielded no significant information, microscopic use-wear analysis currently underway might help determine how the obsidian blades were used.

Discussion and Conclusions

In summary, we have elucidated some strong patterns in our sample of bounded patio groups. The results of our analysis reveal that there is no correlation between houselot size and distance from the site center. We also found no significant relation between the cost of architecture and the distance from the site center. On the other hand, construction cost and size of houselot are correlated with each other, which indicates that the patio groups with the most land were also the ones with the largest, most costly buildings. Given conditions of agricultural scarcity in this part of Yucatán, land for gardening would have been a valuable asset, and many lines of evidence suggest that certain plots within the houselot were artificially fertilized. In patio groups whose buildings exhibit aligned spatial planning, the architecture was more costly, indicating that the corporate groups whose structures exhibited this kind of coherence were able to mobilize more labor. Houselots with sascaberas do not appear any different from houselots without sascaberas, indicating that possession of this economic resource did not contribute to group prosperity. The quantity of metates also fails to correlate with other axes of variation such as houselot size and cost of architecture. Excavations suggest that the amount of metates in a houselot is strongly affected by the length of occupation of that houselot. With regard to other potential economic specializations, horizontal excavations within three patio groups revealed a specialization having to do with obsidian blades in one of the groups.

As a parting comment, our study of bounded patio groups moves us to rethink the reasons why corporate groups form. On the one hand, the definition of a corporate group as "a much more closed and bounded cooperative interaction network than other social units in the community" (Hayden and Cannon 1982:147) seems to match our conception of the bounded patio groups rather well. On the other hand, at Chunchucmil, many of the forces that, according to Hayden and Cannon, bring corporate groups into being—defense, control of trade routes, control of scarce resources—do not appear to apply well to smaller-scale, multi-family groups found in the houselots. For example, at a site-wide scale, defense may have been a concern at Chunchucmil (Dahlin 2000), but it does not seem that the patio groups were defending themselves against each other. Controlling trade routes certainly would have been a concern for the people at Chunchucmil, given that Chunchucmil was heavily involved in local and long-distance commerce, but it is unlikely that each of the approximately 850 patio groups thus far mapped at Chunchucmil formed as part of

an attempt to control a trade route. Finally, control of economic resources would not appear to be a unifying feature at this scale since so
many bounded groups control no apparent resource and those that do
—the sascabera groups—do not exhibit any special patterns. In summary, though we believe that the bounded patio groups do represent
corporate groups, the question of the reasons for forming such groups
remains uncertain. The question merits further research, particularly
at Chunchucmil, whose subsistence strategies and general sociopolitical order are significantly peculiar for a Classic period Maya site.

Part 3 Ancient Politics and Interactions

5

Late Formative and Early Classic Interaction Spheres Reflected in the Megalithic Style

Jennifer P. Mathews and Rubén Maldonado Cárdenas

Interpreting the politics of the Late Formative and Early Classic periods has special challenges. This chapter examines the topic in the Yucatán Peninsula through the concept of the interaction sphere, as reflected in shared architectural styles.

Interaction Spheres

The interaction sphere concept was introduced by Joseph Caldwell (1959, 1964) as a way of coping with changes that occur among societies over broad geographic regions. He notes that as interaction occurs there will be an increase in the rate of innovation, which can result in an identifiable complex of elite material culture. One of the advantages of this concept is that it crosscuts discernible environmental and ethnic areas and instead defines the extent of a population based on interactions and resulting influences and innovations. Caldwell understood that the interactions that were taking place between distinct sociocultural groups were a reciprocal relationship, as well as a necessary force in the evolution of elite institutions among those involved in the relationship (Caldwell 1959:305, 1964:141).

The interaction sphere paradigm suggests that complex, elite social institutions developed out of regional conditions through an informal exchange network among the elites. This model recognizes smaller-scale local networks, but it also acknowledges that interactions were occurring on a regional scale. The need for a regional-scale network would arise out of the demand for distributing scarce but critical resources between the local areas. This distribution would have been handled by the elite class, but it is important to note that their economic monopolization would have been over the dispersion of raw materials and finished products rather than over the actual means of production (Freidel 1979:50).

At least some degree of local interaction and trade is found in most technologically simple societies; however, these phenomena in and of themselves cannot be seen as sufficient conditions for the development of regional interaction spheres and complex society. Instead, a change in the perception of the materials that are considered prestigious would seem to be crucial. By using non-local, non-consumable commodities as forms of wealth rather than local and consumable ones, there is a shift to the reliance on other communities for obtaining them, as well as a need for formalized organization to procure them. This shift, in turn, could lead to the development of specialization in these commodities that would benefit all involved parties. Nonetheless, an intensification of trade in exotic commodities does not necessarily lead to complexity. It would seem that the crucial element would be the consolidation of the local economies into a regional one—an economy in which the commodities that are exchanged become essential to the local economic integration. This necessity for integration should lead to a relationship that continues to solidify this mutual need. If the elites were able to monopolize the interaction occurring between the regional and local networks, a hierarchy of power and status could develop (Freidel 1979:50).

For the Maya lowlands, the model predicts that we should see evidence for the exchange of commodities over reasonably long distances during the Late Formative and Early Classic periods. Evidence has been found for the development of formal trade networks, including warfare and a common conception of what signified an elite good. In addition, large regional centers, which we would expect to develop out of this type of institutionalized interaction, were constructed (Freidel 1979:51).

This interdependence of trade goods and a feeling of belonging to an integrated region would have resulted in a sharing of not only actual trade items and natural resources but ideas and regional styles as well. These collective ideas and styles would be reflected in such items as ceramics, tools, sculpture, religious objects, burial practices, and monumental architecture. In some cases, archaeologists are able to study the emerging patterns to reveal the nature or types of interactions that were occurring (Caldwell 1964:138).

Current Research

In the northern Maya lowlands, work on interaction spheres has focused on the Late Classic, Terminal Classic, and Early Postclassic periods (ca. AD 700–1200). For example, Fernando Robles Castellanos and Anthony Andrews (1986) use the interaction sphere model to argue, on the basis of shared ceramics, architecture, and road systems, that the northern Yucatán Peninsula can be divided into two major cultural spheres during the Classic period: a "western sphere," comprised of north-central Yucatán and the Puuc Hills region; and an "eastern sphere," made up of far eastern Yucatán and northern Quintana Roo. Despite a lack of information for earlier time periods, A. Andrews and Robles Castellanos (1985) suggest that this east-west division for the northern lowlands probably extended back into the Formative and Early Classic periods.

In the Yucatán Peninsula, there are several architectural styles—including Puuc, Rio Bec, and Chenes—that are shared by numerous sites and hence reflect interaction spheres. Despite the plethora of literature dedicated to these three architectural types, few have recognized the significance of the Megalithic style (see, for example, E. Andrews IV and Andrews 1975; Dunning 1992; Pacheco Benítez and Parrilla Albuerne 2004; Roys and Shook 1966; Sidrys 1978; Velázquez Morlet et al. 1991). There are numerous examples of Megalithic architecture across the peninsula, but most are poorly preserved or built over by later architecture, resulting in a general lack of recognition for this architectural tradition or its potential significance.

Although there are clear links during these early periods between Megalithic-style sites such as Aké and Izamal in the western portion of the Yucatán Peninsula, archaeological evidence indicates that this interaction extended to the east as well. This broad sphere of interaction across the northern area is reflected in the remains of numerous cities that shared the Megalithic architectural style (Fedick and Taube 1995; Taube 1995) and in the shared time period in which the Megalithic structures were built, as evidenced by ceramic chronology, corbelled vaults, triadic groupings, and radiocarbon dating (J. Mathews 1998, 2001).

Problems/Questions

As is discussed in Shaw and Johnstone's chapter (this vol.), Classic and Postclassic politics have traditionally been interpreted through such things as the use of texts on monuments, ceramics spheres, burial contents, and shared architectural designs. However, when studying the Late Formative and Early Classic periods in the northern Yucatán Peninsula, researchers face several obstacles to using most of these traditional interpretations. First, unlike most Classic period sites, sites occupied during the Late Formative and Early Classic periods have the unfortunate distinction of being part of the "text-free" zone, as stelae and other art forms with writing are not commonly found this early. Second, large samples of ceramics used for developing ceramic chronologies for these early periods are difficult to obtain for two main reasons: (1) restrictions are placed by the Instituto Nacional de Antropología e Historia (National Institute of Anthropology and History; INAH) on excavation into larger architecture without total consolidation of the structure at great expense; and (2) in the case of Megalithic architecture, the immense stones are challenging and dangerous to excavate. In several cases, this has resulted in smaller ceramic samples obtained from excavation units placed alongside of the architecture. While these samples can certainly be useful, their context limits analysis that can be done on building construction and occupation. This is further complicated by the fact that interpretations of many of the ceramic types are still uncertain. Finally, due to the same limitations for excavating into Megalithic structures mentioned above, for the most part burials and their contents have not been accessed. The result is that we have to rely primarily on shared architectural designs to delineate ancient political interactions in the northern Yucatán Peninsula during the Late Formative and Early Classic periods.

This chapter identifies a northern interaction sphere by first defining what the Megalithic style is. We then outline the distribution of this architectural style for the Yucatán Peninsula by examining all known examples of sites with Megalithic structures. Finally, we will attempt to demonstrate through various dating methods that this style was widespread, primarily during the Late Formative and Early Classic periods, and examine how this reflects the interaction sphere concept.

The Megalithic Style

The Megalithic style is distinctly different in appearance from the other architectural modes within the northern lowlands area (fig. 5.1). As it had only been documented in the western half of the peninsula at sites like Aké and Izamal, the Megalithic style was thought to be limited in its distribution. However, more recently, over twenty-five confirmed sites (and potentially ten more unconfirmed sites) with a pres-

Fig. 5.1 Structure 10 at El Naranjal, an example of the Megalithic style. Note the large stones, rounded corners on the structure, and use of smaller chinking stones. (Photograph by Dominique Rissolo; used with permission.)

ence of Megalithic architecture, including at least twelve sites within the modern border of Quintana Roo, have been documented. Clearly, this would indicate that the style was more widespread than was originally thought, and that it certainly had a presence in the eastern peninsula (J. Mathews 1998, 2003).

Karl Taube (1995) was the first to characterize the architecture in general as having large, well-dressed stones with rounded edges overlaying a rubble core. Many of the blocks of stone are over a meter in length, pillow-shaped, and stacked with roughly broken chinking stones placed in between. Remains of plaster on the facing stones indicate that they were thickly coated to form a smooth exterior surface (Taube 1995:49). At sites such as El Naranjal and Aké, large apron corbels are commonly seen projecting out from the sub-apron wall. Besides being impressive in their size and quality of manufacture, these corbel stones are shaped like a slice of pie, with the narrow end positioned into the rubble-fill interior. The sub-apron wall is at a slight angle, resulting in the exterior side of the apron corbel also being slanted. Some buildings are apsidal in shape (Taube 1995:25); however, apsidal buildings and corbelled aprons are not present at all Megalithic sites.

Placing the Megalithic Style in Time

Associated Ceramic Evidence

As previously mentioned, little excavation has been conducted into
the Megalithic structures at most sites, and therefore few ceramic samples are available for helping to date the architectural style. Nonetheless, with the exception of Structure GT-10 at Ek Balam and Structure
E-VIIa at Chac II, all of the ceramic samples obtained from Megalithic
structures date to between the Late Formative and Early Classic periods. The most prevalent ceramic types include: Clear Slip, Huachinango Bichrome Incised: Huachinango, Ucu Black: Unspecified,
Chancenote Striated: Chiquilá, Tancah Burdo: Tancah, Carolina Bichrome Incised: Carolina, Xanaba, Unto, Laguna Verde Incised: Clear
Slip, and Sierra Red (Boucher 1996; J. Mathews 1998). (See table 5.1 for
breakdown of ceramic types by site.)

Triadic Groupings and Corbelled Vaults

Other ways of dating architecture include architectural associations
such as corbelled vaults and the triadic group formation. When characteristics such as these are found to date consistently to a particular
period, they can be useful in providing additional circumstantial evidence for architectural occupation. Corbelled vaults are common in
the northern Maya lowlands and are generally recognized as an early
trait, usually dating to the Late Formative or Early Classic period (Roys
and Shook 1966:50; von Falkenhausen 1985:129). Evidence of corbelled vaults is known at sites with Megalithic architecture including
Aké, Kantunilkin, El Naranjal, Sihó, and Yaxhom (table 5.1).

Triadic groups are also widespread in the Maya area, including the
major centers such as Calakmul, El Mirador, Cerros, and Edzná. Triadic groups can be defined as a raised rectangular or T-shaped platform that supports three pyramidal structures in a triangular formation. Predominantly, there is a large building in the middle flanked
by two smaller ones on the sides. All three buildings should face into
the middle of the platform, resulting in a central area, and in almost
all instances, a staircase leads to the platform level (J. Mathews 1995).
These triadic groups are also commonly associated with the Late Formative and Early Classic periods (Hansen 1992; J. Mathews 1995). Triadic groups associated with Megalithic architecture are found at the
sites of Aké, Huntichmul, El Naranjal, Site 38, and Yaxuná, providing
additional support for the early dating of the Megalithic style (table
5.1).

Accelerator Mass Spectrometry
(AMS)-Based ^{14}C Dating

In the Maya area, mortar is usually made by burning limestone (Abrams 1994:200–201), potentially leaving behind small pieces of charcoal within the matrix that can be dated through AMS radiocarbon methods. In addition to the direct dating of charcoal inclusions within mortar, the mortar itself can also be dated. At the site of El Naranjal, mortar remains can still be seen on the surface of several Megalithic structures. A total of eleven samples were run on both the charcoal inclusions and the mortar to obtain AMS-based ^{14}C dates that would presumably reflect the construction date of the architecture. Although some samples were contaminated and yielded dates far too old or young, the remaining samples yielded dates that all fell within the range of the Late Formative and Early Classic periods (J. Mathews 2001).

Comparison of Cultural Chronology
for the Megalithic Style

As the associated ceramic chronology and the architectural characteristics of corbelled vaults and triadic groupings indicate a Late Formative to Early Classic occupation, these can be combined with the radiocarbon dates to make a strong argument for a Late Formative and Early Classic occupation of the Megalithic architecture (J. Mathews 1998). As can be seen in table 5.1, the Late Formative through Early Classic occupation of Megalithic architecture is fairly consistent across twenty-three of the twenty-five Megalithic sites with datable materials. As stated above, the exceptions to this early occupation for the Megalithic style are Structure GT-10 at the site of Ek Balam, which has ceramics dating to the late Middle Classic or early Late Classic period, and Structure E-VIIa, dating to the late Terminal Classic period, at Chac II.

Distribution of Megalithic Style Structures across the Yucatán Peninsula

Over thirty-five sites in the Yucatán Peninsula contain known examples of Megalithic-style architecture. The number, size, and preservation of structures, as well as the amount of available chronometric data, vary tremendously from site to site, as a majority of sites have had little more than mapping and test pits conducted. (See fig. 5.2 for

Table 5.1 Comparative Chart Showing the Various Structures at Sites in the Yucatán Peninsula Exhibiting Megalithic-Style Architecture

Site Name	Structure(s)	Megalithic Characteristics	Dating Method	Occupation
Actun Toh	unnamed pyramid	Megalithic stonework	ceramics (Sierra Red, Laguna Verde Incised)	Late Formative–Early Classic
Aké	1, 2, 3, 4, 11, 14, 15, 16, 19, 20, 22, 24	Megalithic stonework, rounded corners, apron corbel, corbelled vaults, stucco masks	corbelled vaults, triadic grouping, ceramics (Saw, Tipikal, and Xanabá Red)	Early Classic
Chac II	E-VIIa	Megalithic stonework	radiocarbon, stratigraphy	Late Terminal Classic
Chan Pich	1, 7	Megalithic stonework	proximity to El Naranjal, shared architectural style	possibly Formative–Early Classic
Cobá	unnamed structure, Xaibe structure	Megalithic stonework, rounded corners	none	unknown
Ciudad Mario Acona	Temples 1, 2	Megalithic stonework, corbelled vaults	similarities to Early Classic architecture, ceramics	possibly Early Classic
Dzonot Aké	V, VI, and IX	Megalithic stonework	ceramics	Late Formative–Early Classic

Site	Structures	Architectural features	Artifacts	Period
Ek Balam	GT-10	Megalithic stonework, rounded corners, apron corbel	ceramics	Middle to Late Classic
El Naranjal	1, 2, 9, 10, 11, 13, 14, 19, 20, 23, Box Ni Group	Megalithic stonework, rounded corners, apron corbel, corbelled vaults, niches, triadic grouping	ceramics (Sierra, Hua-chinango, Ucu, Saban, Carolina), AMS dating, corbelled vault, triadic grouping	Late Formative–Early Classic
Huntichmul	structures B-8, B-9	Megalithic stonework	triadic grouping	Late Formative–Early Classic
Ikil	structure 1	Megalithic stonework, niches	ceramics (Copo complex)	Early Classic
Izamal	Kinich-Kah-Moo, Kabul, Itzam-Na, Hun-Pic-Tok, Chaltunha, and Ppapp-Hol-Chac.	Megalithic stonework, rounded corners, apron corbel, stucco masks	sculptured stucco facades, ceramics (Xanabá Red, Valladolid Bichrome, Shan-gurro Red/Orange, and Polvero Black)	Late Formative–Early Classic
Kantunilkin	structures in Main Plaza	Megalithic stonework	ceramics (Tancah Varie-gated, Tancah Plain, and Chiquila Variegated), cor-belled vaults	Late Formative–Early Classic

Site	Structure/Group	Architecture	Dating evidence	Period
Oxkintok	structure MA-7, Tzat Tun Tzat, DZ-7	Megalithic stonework, corbelled vaults	proximity, similarity to Early Classic structures, overlying structures with Early Classic–Middle Classic ceramics.	possibly Late Formative–Early Classic
Ox Mul	structure 2	Megalithic stonework, corbel apron	ceramics (Carolina, Cetelac, Sierra, Tancah, Balanza, Huachinango, Saban, Timucuy)	main occupation is Late Formative–Early Classic
San Angel	Group A, Group B	Megalithic stonework	ceramics (Saban Coarse, Carolina, Tancah Striated)	possibly Late Formative–Early Classic
San Cosmé	structures 4, 7	Megalithic stonework	proximity, shared architectural style to El Naranjal, ceramics (Sierra Red)	possibly Late Formative–Early Classic
Sihó	structures 15, 16	Megalithic stonework, corbelled vaults	corbelled vaults	Late Formative–Early Classic
Site 38	structure 3	Megalithic stonework, possible triadic grouping	none	too preliminary to assign date
Tres Lagunas	main platform group	Megalithic stonework, rounded corners, apron corbel	ceramics (Carolina, Cetelac, Dzilam, Xanaba, Huachinango, Sierra, Tancah, Timucuy and Saban)	Late Formative–Early Classic

Ucí	structure 2 and one un-named structure	Megalithic stonework	ceramics (Sierra Red, Nolo Red, Ucú Black, Sabán, Polvero Black, Percebes Buff, Unto and Tipikal, Xanabá Red, Shangurro, Timucuy and Dos Arroyos)	Late Formative–Early Classic
Victoria	colonial church, structures 1, 2, 3, 4, 5, Casa de Alux, Acropolis	Megalithic stonework, apron corbel	ceramics (Carolina, Sierra, Tancah, Xanaba, Balanza, Cetelac, Dos Arroyos, Dzilam, Huachinango, Changurro, and Timucuy)	Late Formative–Early Classic
Xcambó	11 structures in main plaza	Megalithic stonework	ceramics	Early Classic
Xcoch	platform and summit platform in site core	Megalithic stonework, corbel apron	none	too preliminary to date
Yaxhom: Nucuchtunich	structure 1, 2, structure 19 stairs	Megalithic stonework, apron corbel, possible stucco masks	ceramics, corbelled vaults	Late Formative–Early Classic
Yaxuná	6F-3, 6F-4	Megalithic stonework, triadic grouping	ceramics, triadic grouping	Early Classic

Note: Comparisons are made between associated ceramic styles, evidence of triadic groupings, and corbelled vaults.

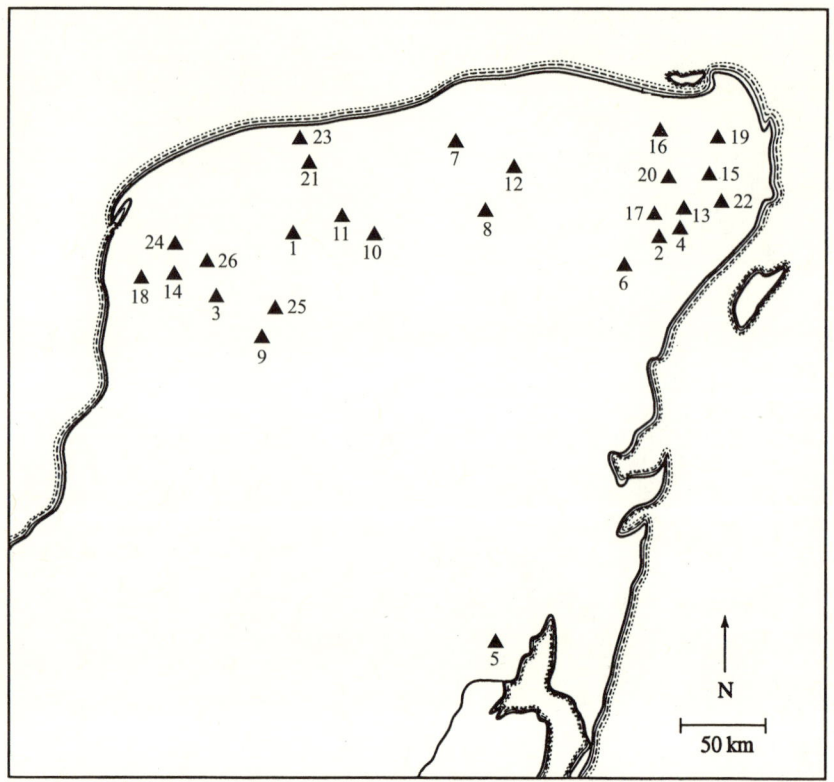

Fig. 5.2 The distribution of Megalithic sites across the Yucatán Peninsula. (1) Aké; (2) Actun Toh; (3) Chac II; (4) Chan Pich; (5) Ciudad Mario Acona; (6) Cobá; (7) Dzonot Aké; (8) Ek Balam; (9) Huntichmul; (10) Ikil; (11) Izamal; (12) Kantunilkin; (13) El Naranjal; (14) Oxkintok; (15) Ox Mul; (16) San Angel; (17) San Cosmé; (18) Sihó; (19) Site 38; (20) Tres Lagunas; (21) Ucí; (22) Victoria;(23) Xcambó; (24) Xcoch; (25) Yaxhom; and (26) Yaxuná. Locations of sites are approximate. (Figure created by Jennifer P. Mathews.)

a map of the location of the sites.) Table 5.1 lists all sites and individual structures with Megalithic-style architecture, the characteristics of the Megalithic style exhibited at the site, and any dating methods, ceramic types, and known dates for the architecture.

The best-preserved and most prevalent examples of the architecture are found at Aké and Izamal in Yucatán state and El Naranjal in Quintana Roo. The site of Aké is located approximately 30 km from Mérida and contains twelve Megalithic structures, as well as settlement with Megalithic platforms. This is one of the most-visited Megalithic sites, due to its proximity to Mérida, its overall size, and the excellent preservation of much of the architecture. Archaeological visitors, including John Lloyd Stephens and Frederick Catherwood (1963), Désiré Charnay (1887), Ralph Roys and Edwin Shook (1966), and Rubén Maldonado (1980, 1981) have all studied and described the ruins.

Although the occupation of Aké was continuous in ancient times, the majority of the structures were originally constructed in the Megalithic style. Ceramics have been obtained from units in the site settlement zone and at the base of the main structures and have been dated to the Early Classic period (AD 300–600; see Quintal Suaste 1993:fig. 1). Additionally, there is evidence of corbelled vaults and triadic groupings, both indicative of a Late Formative/Early Classic occupation (Hansen 1992; J. Mathews 1995).

The site of Izamal covers an area 10 km² and is located approximately 65 km from the modern city of Mérida. This large site was first reported on by John Lloyd Stephens in 1843 (Stephens 1963:298). Later archaeological visitors included Désiré Charnay (1887) and William Henry Holmes (1895). More recent work includes that by Burgos et al. (2003), Kurjack (2003), Lincoln (1980), Maldonado (1990), and Millet and Burgos Villanueva (1998).

Work by Lincoln (1980) found that fourteen of the twenty-three total structures surveyed exhibited Early Classic architectural elements, including Megalithic stonework. Surface ceramics from this area are dominated by Muna Slate, indicating a strong occupation during the late Terminal Classic period, as is also evidenced by the *sacbeob* linking Izamal to the regional sites over which it maintained control. However, excavations into Megalithic platforms recovered Late Formative and Early Classic ceramics (Maldonado 1990).

The site of El Naranjal in Quintana Roo is located southeast of the community of Ignacio Zaragoza in the modern community of Naranjal. This site contains twenty-five monumental structures, of which sixteen exhibit Megalithic stonework (J. Mathews 1998; Pacheco Benítez and Parrilla Albuerne 2004; Taube 1995). The architecture at this site is extremely well-preserved and exhibits only slight architectural modification during the Late Postclassic period (see Lorenzen 1999). Much like at Aké, architecture here exhibits rounded corners, apron corbels, corbelled vaults, and triadic groupings. Dating of the Megalithic architecture at El Naranjal has included ceramic analysis from off-structure units and radiocarbon dating of mortar from the exterior of Megalithic structures dating to the Late Formative and Early Classic periods (J. Mathews 1998, 2001). Most recently, INAH archaeologists excavated and consolidated two Megalithic structures (10 and 14), revealing Late Formative and Early Classic caches of artifacts and ceramics, confirming earlier research (Pacheco Benítez and Parrilla Albuerne 2004).

As these three sites are some of the largest in their respective regions, exhibit the most Megalithic structures by far over all other sites, and seem to have been solidly established by the Late Formative and Early Classic periods, they would logically be the models whose architectural style was emulated by other sites. Megalithic architecture is evidenced at twenty-five other sites across the peninsula, although the majority of these sites contain three or fewer structures constructed in the large-block style. Of these twenty-five sites, approximately fifteen of them have at least tentative evidence of Megalithic architecture dating to the Late Formative or Early Classic period. These include Actun Toh, Ciudad Mario Acona, Dzonot Aké, Ikil, Kantunilkin, Oxkintok, Ox Mul, San Angel, San Cosmé, Tres Lagunas, Ucí, Victoria, Xcambó, Yaxhom, and Yaxuná (see fig. 5.2 for locations).

The Cave of the Mot Mot, or Actun Toh, is located approximately 1 km from the community of San Juan de Dios, Quintana Roo and is located within walking distance of the sites of San Cosmé and El Naranjal. The main chamber of the cave is 40–50 m in diameter and 6 m high. Directly beneath the entrance hole in the ceiling is a large terraced mound approximately 4 m high. While the mound undoubtedly was formed from the overlying cave entrance, it has been modified extensively. The majority of the stones are carved, and the overall construction of the mound is reminiscent of the Megalithic style. There are two possible stairways on either side of the structure, and at least four moderately well-preserved terrace risers are visible on its western slope. However, despite its location within the cave chamber, the mound is for the most part in a severe state of disrepair (Rissolo 2003:38). The pyramid and the rest of the cave are littered with cultural materials, including altars, ceramics, and metates. It is difficult to associate the surface artifacts with the construction of the pyramid because the cave was probably used as a pilgrimage site from the Late Formative period well into the Late Postclassic period. However, a quantitative overview of the ceramics obtained indicates that primary usage of the cave occurred during the Late Formative and Early Classic periods, and Rissolo feels the structure was initiated by this time (Dominique Rissolo 2003:54, personal communication 2004).

Located in the far south of Quintana Roo, the site of Ciudad Mario Acona is approximately 29 km southeast of the modern town of Bacalar. It covers a 700 m area north to south and consists of three small nucleated centers made up of platforms and temples. Of interest here are Temples 1 and 2 of the first center. Both structures are badly collapsed; however, Lizardi Ramos (1940) reports tall corbelled arches and moldings with very large blocks up to .90 m long. Looking at the photos of the original site report, it would also appear that Temple 2 had rounded corners (see Lizardi Ramos 1940:photo 6). Excavations were conducted into Temples 1 and 2, but the ceramic types were not presented in the 1940 report. Lizardi Ramos did not feel comfortable dating the site due to lack of Long Count dates or monuments; however, he said architectural comparisons with the temples of Tikal would indicate an occupation during the Early Classic period (Lizardi Ramos 1940). If these structures are in fact constructed with Megalithic block masonry, they would represent the southernmost examples of the Megalithic style.

The ruins of Dzonot Aké are located approximately 20 km east of the modern city of Tizimin, Yucatán and 1 km outside the modern settlement of Dzonot Aké. The site was previously mapped by David Webster in the 1970s, and his report indicates that there were possible Megalithic structures including: Structure I, a large pyramid approximately 21 m high; Structure III, a low terraced mound; and Structure V, a range platform (Webster 1979:Map 4). Unfortunately, most of the larger stonework has been looted in recent years. Although few Megalithic blocks remain, the construction style of the range platform is reminiscent of structures at El Naranjal and Aké. Although this information certainly does not demonstrate that the structures were built in the Megalithic style, a few remaining stones (measuring approximately 80 × 40 cm and 80 × 55 cm, respectively) and similar construction styles do indicate the distinct possibility. Ceramics excavated from the site fall into the Late Formative and Early Classic periods (Webster 1979).

One of the few sites deemed in the archaeological literature to have architecture with Megalithic stonework, Ikil is located 26 km southeast of Chichén Itzá, near the modern town of Yaxcaba in Yucatán state (E. Andrews IV and Stuart 1968). Structure 1 is a platform with three levels, capped with a summit temple, reaching a total of 25 m in height. This summit temple has outer walls that are lined with well-dressed, horizontally laid Megalithic blocks up to 1.5 m long, .8 m wide, and .4 m high. Like many other Megalithic-style buildings, the corners of the temple summit were rounded (E. Andrews IV and Stuart 1968:75). Around the main doorway there are recessed niches that probably acted as panels for tenoned sculptures (E. Andrews IV and Stuart 1968:73). These same features are seen on Structure 1 at El Naranjal (Taube 1995:26). Based on surface ceramics (all of which were included in the Copo complex), the authors estimate the architecture to date to the Early Classic period (E. Andrews IV and Stuart 1968:75).

However, closer observation reveals that this building is slightly different from other Megalithic buildings. These differences include: (1) the large blocks are almost completely rectangular, as opposed to the pillow-shaped blocks known at sites like El Naranjal and Aké; (2) due to their rectangular shape, the stones were closely fitted and required a minimum of chinking stones placed in between for stabilization; and (3) the vaults found in association with the Megalithic construction were not truly corbelled like those at El Naranjal and Aké.

The ancient ruins of Kantunilkin in Quintana Roo are poorly preserved and heavily looted. Outside of the original report by William Sanders in 1960, there is little information available on the site. Fedick and Taube (1995) noted the remains of two large pyramids in the center of town, in addition to well-dressed Megalithic blocks in a platform adjacent to a modern plaza close to the entrance of the modern town. Small cobbles line the outside of the pyramid structures, and with the exception of one well-dressed stone on top of the largest building, any evidence of sizable facing stones has disappeared. A local archaeologist explained that the stones have been used to build property walls and were mined by a local cement company to be ground up into powder for building materials (Antonio Centeno Mena, personal communication 1997).

Dating for Kantunilkin is preliminary; however, Sanders reports that the ceramics of the area include Tancah Variegated, Tancah Plain, and Chiquila Variegated, dating to the Late Formative and Early Classic periods (Sanders 1960). In addition, Fedick and Taube (1995) noted Early Classic basal flange polychrome vessel sherds associated with the main platform.

The center of Oxkintok in western Yucatán, with carved monuments and beautiful Puuc-style architecture, represents one of the earliest major centers in the northern Maya lowlands (Rivera Dorado 1987, 1989). Much of the architecture at this site exhibits characteristics known at early Maya centers, including rounded corners, corbelled vaults, and some use of Megalithic stonework. Structure MA-7, a pyramidal structure on the Grupo May platform, has rounded corners. Still another example of possible Megalithic architecture is the Tzat Tun Tzat structure, also known as the Labyrinth. It is a 7 m high mound with a series of complicated passageways connecting small rooms. The roofs of these passageways are of stepped corbelled vault construction with large well-dressed stones and remains of chinking stones, simi-

lar to those known at El Naranjal and Aké. Furthermore, El Dzib group
contains a structure known as DZ-7, which has a stairway on the north
end of the building constructed of massive blocks.

All of these structures exhibiting these early characteristics are
thought to date to the Early Classic period. Although ceramics have not
been obtained for Structure MA-7, the style of the architecture as well as
its close proximity to the Early Classic Structure MA-1 suggest an Early
Classic date (Rivera Dorado 1989:88). For the Tzat Tun Tzat structure,
the chronology is complicated, but it is believed to have been occupied
from the Late Formative period to the Early Postclassic period. Underlying constructions and burials such as Tomb 1 (containing Early Classic and Middle Classic ceramic vessels) have caused some to hypothe-
size that the original design of the labyrinth was constructed in the
Late Formative period and elaborated on in the Early Classic period
(Ardren 1997:193–94; Rivera Dorado 1989). Finally, although the Dzib
group (and hence Structure DZ-7) remain unexcavated, their similarity
to the Early Classic Xhanha group at Yaxuná suggests an Early Classic
date (Ardren 1997:196).

The site of Ox Mul was first documented by Glover and Esteban
Amador in 2001. Located approximately 75 km northeast of El Naranjal, the closest modern community to the ruins is the town of Francisco May about 6.5 km southwest. The site consists of two main groups
known as Structures 1 and 2. Structure 2 is approximately 75 m by 45 m,
is oriented east-west, and contains well-preserved Megalithic architecture on the north side of the platform. Two superstructures on the eastern side of platform contain possible corbelled aprons as well. Ceramics were surface collected as well as obtained from test pits placed next
to Structure 2. While the ceramic material ranges from the Middle Preclassic period to the Late Postclassic period, Megalithic constructions
seem to be associated with Late Preclassic and Early Classic ceramics such as: Carolina, Cetelac, Sierra, Tancah, Balanza, Huachinango,
Saban, and Timucuy (Glover and Esteban Amador 2002, 2004; Fabio
Esteban Amador, personal communication 2004).

The site of San Angel is located north of Kantunilkin in northern
Quintana Roo. Although known by 1984, Taube and Gallareta Negrón
(1989) first recorded the mural paintings and standing architecture
in detail as part of the San Angel Survey Project. The ruins include
two plaza groups located approximately 3 km apart. Group A, closest
to the modern town of San Angel, is a rectangular platform measuring approximately 45 m by 58 m, supporting nine structures. Nearby,
Group B is topped by six structures. While this site is best known for its
Late Postclassic murals, surface ceramics collected from both groups

indicate that the greatest level of occupation was during the Late Formative and Early Classic periods, as evidenced by Saban Coarse, Carolina, and Tancah Striated ceramics (Gallareta Negrón and Taube 2005; Taube and Gallareta Negrón 1989). Although Taube and Gallareta Negrón describe the architecture as poorly preserved and with little remaining evidence of Megalithic facing stones, it is believed that the original architecture was in the Megalithic style. On Group A, the Late Postclassic walls are made up of small, poorly dressed stones with a few well-dressed stones just less than a meter in length lain on top. It is possible that these larger stones were reused from an earlier occupation.

The small center of San Cosmé in Quintana Roo is directly linked to El Naranjal by a 3 km long sacbe but interestingly shows little Megalithic stonework. Structures 4 and 7 do show some evidence of large blocks, and at one time many of the structures may have exhibited the well-dressed stonework known for the area. Much of the stonework may have been lost to modern looting; however, as San Cosmé is a fairly minor center, it may have had only a limited amount of the more impressive stonework. Nonetheless, the direct connection to El Naranjal and the presence of Late Formative Sierra Red ceramics at the site would indicate that the Megalithic style was in existence, even if only in a limited quantity (Taube 1995:49).

Although the site of Tres Lagunas in Quintana Roo is well-known by locals, it was not officially known to archaeologists until 1993 when it was mapped by Fedick and Hovey (Taube 1995). The structures are located 4 km east of the modern community of El Cedral, just south of Kantunilkin. The site consists of a massive basal platform located along a series of lakes. The platform is over 2 m high and covers an area roughly 120 m × 130 m. There is an adjoining platform that projects from the northeastern corner, between two of the lakes. The sides of this tall platform are well preserved, and the largest Megalithic stones, just less than 2 m in length, are found at the upper end of the vertical face. The northern and eastern sides of the platform exhibit rounded corners (Taube 1995:49).

The main building platform supports a complex of five smaller superstructures. The lower portion of the largest of these abuts directly on the northern basal platform edge, creating a single vertical face. The stairway of this principal building platform is located toward the eastern side. The smaller northern basal platform also sustains a series of structures on its west, north, and south sides. The northern structure of this plaza group is especially well preserved and is clearly faced with Megalithic stonework (Taube 1995:49, fig. 2.25). Ceramics have been obtained from associated test pits by Jeffrey Glover and

Fabio Esteban Amador. Although occupation ranges from the Middle Formative period to the Late Postclassic period, the main occupation appears to date to the Late Formative and Early Classic periods. Ceramic groups found include Carolina, Cetelac, Dzilam, Xanaba, Huachinango, Sierra, Tancah, Timucuy, and Saban (Fabio Esteban Amador, personal communication 2004).

The site of Ucí is located in the middle of the modern town of Ucí, approximately 2 km northeast from Motul, Yucatán. Three structures (2, 3, and 12) still exhibit Megalithic architecture, although heavily damaged and looted of much of the original stones (Kurjack 2003: 11; Maldonado 1980, 1995:74, 77). In general, Ucí presents a long sequence of development that lasted from the Late Formative period to the Postclassic period. According to Maldonado, however, the high concentration of ceramics indicates that the site developed in the Late Formative period (represented by the ceramic groups Sierra Red, Nolo Red, Ucú Black, Sabán, Polvero Black, Percebes Buff, Unto, and Tipikal) and culminated in the Early Classic period (represented by the groups Xanabá Red, Shangurro, Timucuy, and Dos Arroyos) and then went into subsequent decline.

The ruins of Victoria are located in northern Quintana Roo, off of Highway 180 and southeast of the modern community of Agua Azul. J. Mathews (2003) first became aware of this site in 1996, when local consultants in the village of Naranjal reported that there were ruins near the village of Victoria similar in appearance to those at El Naranjal. Not only do the two sites share an architectural style, but El Naranjal is 22 km directly east of Victoria. Since that time, Jeffrey Glover and Fabio Esteban Amador have located additional monumental structures at Victoria (Glover and Estaban Amador 2004). Those with Megalithic architecture include five residential structures (Structures 1, 2, 3, 4, and the Casa de Alux), a late colonial church built out of looted Megalithic stones (J. Mathews 2003), and an Acropolis (Glover and Estaban Amador 2004).

The Acropolis is a major platform covering 18,000 m² and supporting ten structures. Structure 1b has its own Megalithic superstructure: Structure 1b-1, which is 25 m by 12 m and includes massive stones and evidence of a corbelled apron. Additionally, walls on the western and eastern sides of the platform have 2 to 3 courses of Megalithic stonework, and Structure 1g in the southwest corner of the platform contains the best-preserved Megalithic architecture at the site (Glover and Esteban Amador 2004). Ceramics have been obtained from associated test pits, and although the occupation ranges from the Middle Formative period to the late Terminal Classic period, the overwhelming majority of ceramics date to the Late Formative and Early Classic periods. Ceramic types include Carolina, Sierra, Tancah, Xanaba, Balanza, Cetelac, Dos Arroyos, Dzilam, Huachinango, Changurro, and Timucuy.

Located on the north coast of Yucatán state and to the north of Aké, Xcambó was a major port for trade during the Early Classic period (Sierra Sosa 1999, 2001:27). Although the site is relatively small, measuring roughly 150 m × 750 m with a total of eleven monumental structures in its main plaza, the architecture combines Petén-style constructions with Megalithic stonework. Although many of these "megaliths" seem to be smaller than those used at sites such as Aké and El Naranjal, the architecture nonetheless appears to incorporate large stonework and rounded corners. Sierra Sosa hypothesizes that Xcambó was under the control of nearby Izamal and likely emulated their architectural style in all of the monumental architecture. Based on the associated ceramics, Sierra Sosa dates the occupation of Xcambó to the Early Classic period (AD 250–600) (Sierra Sosa 1999).

The site of Yaxhom is located southwest of Loltún Cave in western Yucatán state and includes the associated groupings of Nucuchtunich and Nohoch Cep. Although the site center still includes several examples of Megalithic stonework, much of the architecture has been destroyed due to agricultural development and the looting of stonework. Structure 1 of the associated Nucuchtunich group is a small pyramid representing the best example of intact Megalithic construction at the site. This structure has a fairly well-preserved platform with large stones lined with chinking stones, an apron corbel, and an unusual Megalithic superstructure. This upper structure is a multi-chambered construction with an apron corbel. A footing stone for a mask, possibly covered in stucco, like those known at Izamal (Stephens and Catherwood 1963:Plate LI) and possibly Aké (Roys and Shook 1966:49), is evident. Other constructions include Structure 2 (a poorly preserved Megalithic structure) and Structure 1 (an 8 m high pyramid constructed in the Megalithic style) at the Nohoch Cep group (Dunning 1992:180–81).

Structures at the Main group at Yaxom also exhibit Megalithic characteristics, including: Structure 2, a poorly preserved range structure on top of a Megalithic platform; Structure 18, a two-room vaulted building built on a Megalithic-style basement terrace; and Structure 19, a tall platform with a Megalithic stairway (Dunning 1992:180–81). A small surface collection made at Nucuchtunich contained mostly Muna variants and Cehpech wares, which extend back into the Early Classic period (Nicholas Dunning, personal communication 1998).

The large center of Yaxuná is located 25 km south of the site of Chichén Itzá and is directly connected by a 100 km long sacbe to the site of Cobá. The Yaxuná Archaeological Project collected a considerable amount of information on the site history and chronology (see, for example, Ardren 1997; Freidel 1986; Shaw and Johnstone 2001; Stanton 2000; Suhler 1996). Ceramic evidence indicates that the site was originally settled in the Middle Formative period, monumental architecture was constructed during the Late Formative period, and an increase in scale and caliber of the architecture occurred during the Early Classic period (Ardren 1997:148). Yaxuná was heavily involved in the politics of the region, as evidenced by royal burials, foreign-influenced ceramics, long-distance road systems, and trade goods (Ardren 1997:16). Much of the later architecture of the site is constructed in the Puuc style, with elaborate colonettes and sculpture.

However, Megalithic construction is seen in Structures 6E-12 and 6E-14, which are located near the exact center of Yaxuná and the main temple complex. Neither of the structures has been excavated, and therefore, no associated ceramics are available (Traci Ardren, personal communication 1998). However, there are Megalithic-style stairways on two buildings—Structures 6F-3 and 6F-4 of the North Acropolis—both of which have been subjected to extensive excavation. Structure 6F-3 is a 16.5 m high pyramid on the northern end of the triadic North Acropolis, while Structure 6F-4, an 8 m tall pyramid, is on the eastern end. Although it was elaborated upon in later, overlying constructions, this triadic arrangement was believed to have been established by the Late Formative period (Suhler 1996:162). Ceramic chronology indicates that Structure 6F-3 was occupied from the Early Classic period (AD 250/300) through the Postclassic period.

Several additional sites exhibit Megalithic-style architecture or stonework, although work on these structures is too preliminary to assign dates. These include the sites of: Chan Pich, Quintana Roo, Structures 1 and 7 (Rissolo 1997:17); Cobá, Quintana Roo, the Xaibe structure and an unnamed structure at the junction of Sacbe 1 and 3 (Folan et al. 1983:75, 83, 223; Navarrete et al. 1979:53); Huntichmul, Yucatán, Structure 9 of Group B (Dunning 1992:231); Sihó, Yucatán, Structures 15, 16 and 5D1 (Dunning 1992; Fernández Souza et al. 2002); Site 38, Quintana Roo, Structure 3 (Glover and Esteban Amador 2004); and Xcoch, Yucatán, an unnamed platform in the site center (Dunning 1992:171).

In addition to the above-named structures, an article by Velázquez Morlet et al. (1991) mentions several sites in northeastern Yucatán that have architecture exhibiting possible Megalithic characteristics, comparing them to Ek Balam and Aké. They mention several sites in the

Chikinchel province of Yucatán state that have large platforms constructed with huge blocks of stone at sites such as Dzibalkú, El Sauce, Ichmul II, Nuevo León, San Fernando, San Miguel, San Pastor, and Xpoop. The Chikinchel province is generally associated with the Late Postclassic period; however, the authors note that the Megalithic elements probably date to the Early Classic period (Velázquez Morlet et al. 1991). In addition, Nicholas Dunning (1992) notes other structures within the Puuc area that may also fall within the realm of the Megalithic style. These include Structure 6 of the East Valley group of the site of Yaxché Xlapak and Structure 3, an oblong pyramidal platform with a stairway on the eastern side that is constructed of Megalithic blocks, at the Cab site (Dunning 1992:262).

Two additional sites, Ek Balam and Chac II, each exhibit one Megalithic-style structure that does not fall into the Late Formative/ Early Classic pattern prevalent in the rest of the peninsula. Ek Balam, located in Yucatán state, is approximately 51 km northeast of Chichén Itzá and 60 km northwest of Cobá. The site center includes one Megalithic building—Structure GT-10—a tall platform constructed entirely of massive, horizontally laid, well-dressed stones up to fifteen courses high. Lined along the top of the structure are large corbel stones, all rectangular-shaped, with the exception of the wedge-shaped corner stones.

GT-10 was reconstructed in 1994 by INAH archaeologists, who interpret it as a Late Classic structure with earlier construction phases, based on test pits placed next to Postclassic superstructures (Vargas de la Peña et al. 1994). Bey has re-evaluated ceramic sherds collected during the 1986 field season from a test pit located in the center of the superstructure below the floor of the main platform. The fill contained evidence of Muna Slate and Chumayel Red on Slate fragments (dating to the Late Classic period) mixed with Early Classic and Formative periods material. This would indicate that this Megalithic structure was constructed after the production and use of these slate wares, placing it later than nearly all other known Megalithic constructions. Bey points out that these slate wares first appear sometime during the Early Classic period, but that these sherds in particular look like they actually date to the Middle Classic to early Late Classic periods. Although Bey and the authors feel that the architectural style contradicts these findings, an Early Classic date cannot be argued for Structure GT-10 at this time. However, it should be noted that recent INAH excavations have revealed a large amount of Early Classic pottery, including eighty Huachinango vessels found near the ballcourt, which would indicate a strong Early Classic presence at the site (George Bey, personal communication 2004).

Located in the Puuc region near Kabah and Sayil, the site of Chac II first appears during the Early Classic period, expands during the Middle Classic period (AD 500–650), and reaches its apogee by the Late Classic period (AD 650–800) (Smyth 1998; Smyth et al. 1998). This major center has an unusual example of Megalithic-style architecture. The Great Pyramid (Structure E-VII) was first constructed in the Early Classic period around AD 400 and then was overlaid with a construction phase that appears to have foreign influences and a later Puuc-style occupation. This was followed by a Megalithic-style construction (E-VIIa) that Smyth dates to the late Terminal Classic period (see Smyth and Ortegón Zapata, this vol.). He has also found a Megalithic stairway in the construction of the Intermediate Pyramid substructure. Although Smyth does agree that most examples of the Megalithic style correspond to the Early Classic period, he feels that this particular example should be assigned this late date on the basis of ^{14}C and stratigraphic dating (Michael Smyth, personal communication 2004).

We believe these two examples of Megalithic structures are anomalies in the general Late Formative and Early Classic pattern. It is likely that they represent examples in which the Megalithic style extended into the Middle Classic or Late Classic period at Ek Balam (Bey et al. 1997:239) and the Terminal Classic period at Chac II (Smyth and Zapata Ortegón, this vol.). It would seem that if these two centers were representative of the major Megalithic sites, there would be more structures exhibiting the large-block construction and they would date to roughly the same time period as each other. Instead, one dates to the Late Classic period while the other dates to the late Terminal Classic period. While these two structures should not be discounted in our understanding of the Megalithic style, what they may actually signify are separate attempts to reflect back upon earlier centers such as Aké, Izamal, or El Naranjal and their former power, much like many of the Late Postclassic sites in this region (see, for example, Lorenzen 1999).

Thus, despite a few exceptions, the widespread examples of Megalithic-style architecture found in the northern Yucatán Peninsula date to the Late Formative and Early Classic transition periods. This shared architecture found in a northern sphere across the peninsula is the physical remnant of what was once a link between sites. This regional network would have allowed for the distribution of the concept of the Megalithic style, as well as a sharing of resources and goods.

Discussion and Conclusions

The interaction sphere model has proven useful in interpreting the politics of the northern Maya lowlands during the Classic and Postclassic periods. In this chapter we have set out to provide the theoretical orientation of the interaction sphere concept; define the Megalithic architectural style including its geographic extent, architectural characteristics, and temporal employment of the style; and provide evidence to argue for an ancient Maya interaction sphere that spanned across the northern Yucatán Peninsula during the Late Formative and Early Classic periods, rather than the divided east-west distribution known for later periods.

However, dating northern Maya lowland sites with Late Formative and Early Classic components poses challenges that researchers working during later periods or in the south do not face. These include a lack of monumental texts, the difficulty of obtaining ceramic samples or burials from within larger structures, and the inability to date architecture directly through ceramic chronologies. This means that to use the interaction sphere paradigm, we are often limited to using evidence from the surface, such as the shared architectural style, corbelled vaults, triadic groups, associated ceramics, and AMS dating of mortar. Nonetheless, this combination of methods and the recent INAH excavations into Structures 10 and 14 at El Naranjal argues powerfully for a consistent occupation period for Megalithic style, indicating a northern interaction sphere during the Late Formative and Early Classic periods in the Yucatán Peninsula.

Finally, despite the Megalithic style being widespread across the Yucatán Peninsula, it has not yet been recognized as one of the hallmark styles like Puuc, Rio Bec, and Chenes. Evidence presented here indicates that it does warrant being identified as a regional architectural style for the peninsula. Continuing work on this issue will allow for new insights into the political, ideological, and cultural landscape of the ancient Maya in the Yucatán Peninsula for the Late Formative and Early Classic periods.

6

Foreign Lords and Early Classic Interaction at Chac II, Yucatán

Michael P. Smyth and David Ortegón Zapata

Highland-lowland interaction in the Early Classic period long has been a subject of intense debate in Mesoamerican archaeology. This debate has become reinvigorated recently because of epigraphic decipherment suggesting an active takeover of the Maya lowland centers of Tikal and Copán by people closely affiliated with highland Teotihuacan (Sharer 2003; Stuart 2000). Shifting their emphasis to the point of polarization, some researchers argue for outright Teotihuacan domination of Maya centers (Cowgill 2003; Sanders and Price 1968). Others see influence as more covert in nature, involving selective appropriation, emulation, and manipulation of foreign imagery by Maya elite for local status enhancement (Braswell 2003; Demarest and Foias 1993). With little safe middle ground in this debate, it is sometimes overlooked that almost all of Mesoamerica was an interacting culture area since Formative times. The culture dynamics of interaction, particularly during the Early Classic period, must have differed widely from region to region and from site to site. Considering how these complex relationships must have evolved, changed, and dissolved as events, circumstances, and processes warranted forces a sobering realization that there had to be great variability of foreign interaction. Across the Maya area, this range of variability is too poorly understood to simply assert takeover versus emulation scenarios: these opposing perspectives are of limited utility in assessing the meaning of this fundamental archaeological issue. An examination of early foreign interaction at the Puuc Hills center of Chac II (Chac) will help redress the issue of variability and provide a new perspective from a region long thought to have been outside the sphere of Teotihuacan influence.

Current Research

Nine seasons of research at the Maya center of Chac are leading to a reconsideration of the cultural processes of foreign interaction at a time before the great Terminal Classic "florescence" (fig. I.1). The work at Chac supports two important conclusions: (1) the Puuc region was occupied with significant settlements in the latter part of the Early Classic period (AD 300–600); and (2) there were intense interactions with people outside the Maya area, including the presence of foreigners affiliated with the central Mexican metropolis of Teotihuacan. These data have far-reaching implications for Mesoamerican archaeology and can open a whole new vista into a relatively unexplored chapter of northern Yucatecan prehistory.

This chapter will present evidence for a foreign elite presence at Chac near the end of the Early Classic period. It has been argued elsewhere that spatial, contextual, and stylistic patterns of residential architecture, mortuary customs, and domestic assemblage support a foreign presence of non-elites, perhaps belonging to a small group of long-distance traders who married into local Maya populations (Smyth and Rogart 2004). Foreign icons and iconography from the Great Pyramid Plaza recovered during the 1996–2001 field seasons also suggest that elite emulation was a factor from the Early Classic to Late Classic periods (Smyth et al. 1998; Smyth and Rogart 2004). The behavioral contexts of architecture, artifacts, and iconography recovered from the Grecas Plaza during the 2002 and 2003 seasons, however, provide compelling new evidence that goes beyond symbolic emulation to support a case for an actual presence of foreign elite. The meanings of such high-level social interactions are explored and a tentative explanation is offered, addressing why the site of Chac in particular and the Puuc region in general were of strategic interest to outsiders at this time.

The Chac II Project: 1995–2001

The field research at Chac began as an outgrowth of the surface collection survey at Sayil (Smyth and Dore 1992, 1994; Smyth et al. 1995). A three-story palace, large pyramid, and two partially standing stone buildings at Chac were first thought to be settlement outliers of Sayil. In 1995, a program of intensive survey undertaken at Sayil began at Chac including settlement mapping, systematic surface collection, and soil testing. It became immediately apparent that Chac was not an outlier but rather an earlier independent settlement, as evidenced by a vacant intersite zone and almost exclusive presence of early-style architecture. The survey documented a surprisingly dense settlement area covering 3 km². By 1996, it was decided that to reconstruct Chac's chronology and occupation would require large-scale architectural excavation of selected monumental edifices. The Chac Pyramid Plaza (Gran Plaza) was chosen because it contained the largest pyramid at the site, was attached to a plaza surrounded by a number of stone buildings seemingly in various architectural styles, and appeared to be one of the oldest continuously occupied parts of the site. It was here that we hoped to find architectural stratigraphy, carbon samples for radiocarbon dating, and both complete vessels and potsherds in sealed stratigraphic deposits. A program of test pitting was also carried out at architectural and non-architectural contexts across the site, including an agricultural terrace in the north (Smyth et al. 1998).

Architectural excavation, consolidation, and restoration, where possible, took place at the Gran Plaza from 1996 to 2001. As many as five separate construction phases were uncovered on the Great Pyramid. The Phase I pyramid substructure named Ka'nah was radiocarbon dated to AD 370 ± 60 and shows unusual characteristics such as small facing stones used as staircase treads, risers, and foundation walls; balustrades; and evidence for a vaulted roof temple building at the summit (table 6.1). Significantly, the Phase III pyramid, dated to AD 520 ± 40, was constructed in a foreign style with adobe-like stonework, a thin layer of hard stucco, and a Megalithic staircase. Evidence shows that this unusual pyramid and summit building was embellished with stone sculpture, painted stucco decoration incorporating Teotihuacan-like icons, symbolism, and polychrome painting. Phase IV is identified as the native Early Puuc style and was confined to the south face of the pyramid, featuring multiple sun-god stucco masks and a wide staircase of stone blocks ascending nine tiers leading to a three-room vaulted temple building with colonette decoration. The final Phase V dates to the Late Classic/Terminal Classic period and is Megalithic-like in style with "pillow-shaped" slab stones, chinking stones, and heavy coats of stucco covering Phase III to form an apsidal-shaped base on three sides and a large stucco mask of a feathered serpent on the southeast. It is near the end of this phase that all plaza buildings were physically destroyed and the plaza itself was ritually terminated by the construction of wall segments closing off all points of access (Smyth 2002; Smyth and Rogart 2004).

Seven plaza buildings (E-II, E-III, E-IV, E-V, E-VI, E-VIIa, and E-VIIb), two ramps, the Great Pyramid, an attached frontal platform, and a summit temple (E-I) together form a pentagonal-shaped plaza. Within

Table 6.1 Selected Radiocarbon Dates from Chac (II), Yucatán

| Field Specimen | Year | Lab No. | Conventional Date AD | | | Context |
			C-14 Age BP	Calendar Date AD	Calibrated C-14 AD[a]	
30511	1996	Beta-98318[b]	1190±100	760	655–1025	Grecas chultun; above-floor stratum
30513	1996	Beta-98319	1610±60	340	340–600	Grecas chultun; embedded in-floor
30539	1996	Beta-98322	1250±60	700	665–905 and 920–950	E-VIIa, uppermost stucco floor
30545	1996	Beta-98323	1430±60	520	540–690	E-VIIa, megalithic platform floor
30533	1997	Beta-114546	1330±50	620	640–790	central altar; within vessel offering
30711	1997	Beta-114547	1250±50	700	670–890	E-VIIa, upper stucco floor
30713	1997	Beta-114548	1330±50	620	640–790	E-VIIa, lower stucco floor
30730	1997	Beta-114552	1580±60	370	380–620	N Pyramid Plaza lower surface
40001	2000	Beta-148714	1430±40	520	560–670	pyramid trench; chultunera offering

[a] All dates were calculated using the Accelerator Mass Spectrometer (AMS) or Standard Radiometric techniques.
[b] 2 sigma, 96% probability.

the plaza center are a circular stone altar and two column stones, with another altar and column to the northwest. Although all visible buildings were in the Early Puuc style, there appear to be early and late phases of this style. In addition, all of the early-phase buildings contained substructures in the form of Megalithic-like platforms (E-II, E-VIIa, and E-VIIb), Proto-Puuc-style platforms (E-V), and a substructure ramp of roughly shaped stonework set with mud mortar (South Ramp). These architectural data clearly show that there was an enclosed Pyramid Plaza in this space since the construction of the Phase III pyramid in the sixth century, if not earlier.

Two non-elite residential platforms, the Sacta group and the Platform group, yielded some of the more significant results of the project. The objective at these groups, located to the west and northwest of the Gran Plaza, was to document patterns related to domestic activity both early and late. We were also drawn to these groups because surface mapping showed unusual spatial arrangement and orientation for numerous perishable buildings located upon different surface levels. Both residential groups contained typical foundation braces for perishable buildings dated to the Late Classic period but no stone-walled or roofed buildings. Immediately below Late Classic constructions were boulder-stone foundations revealing large modular-style, multi-room substructures integrated with corridors and walled interior patios. These substructures were oriented about fifteen degrees east of north and show spatial conventions and organization similar to central Mexican domestic architecture at this time, especially apartment compounds found at Teotihuacan. Located on a high hill, the Sacta group is better preserved and shows these spatial-residential characteristics as well as a square altar of rough stone masonry, three modular room blocks connecting onto interior patios, and evidence for a heavy enclosure wall surrounding much of the compound. This group was filled in and used as a base for the Late Classic construction of a two-room building and a bare platform.

A non-elite foreign presence at the Sacta and Platform residential groups is also supported by domestic artifact and mortuary patterns. A total of twenty-three human burials were found within subfloor contexts. All burials were in seated or flexed positions placed into circular stone-lined cists or between pottery vessels in tight fetal positions. The latter were infant-perinatal burials including five interred apparently as ritual offerings within a large stuccoed depression below the east wing of the Sacta substructure. These "round" burials and child offerings show striking similarities to central Mexican mortuary customs during the Middle Classic period.

Artifact assemblages also support the hypothesis that the Sacta and Platform groups were foreign-occupied. Early-style and foreign-style vessel forms and decoration such as slatewares; triple-handle water jars; miniature vessels including *candelero, venenera,* and *chultunera* forms; as well as ring-base and Thin Orange ware bowls were recovered in burials and caches. Lithic patterns revealed a high percentage of obsidian (some of highland Mexican origin and workmanship) and many thin biface projectiles identified as atlatl dart points (Smyth and Rogart 2004). There were numerous worked and perforated marine shells including large bivalve shells for a necklace and pyrite mirror fragments often considered to be central Mexican-related status items and costume accessories (Stone 1989:157).

The Grecas Plaza: 2002–2003

The Grecas Plaza is a likely setting for the residence of a foreign elite at Chac. Indeed, the Grecas Plaza, although compact and comparatively small, is set in the middle of the Central Acropolis, the most massive construction complex at the site and one of the largest in the Puuc region. The Plaza is demonstrably early, being at the lowest known surface level of the Acropolis, radiocarbon dated to AD 340 ± 60 by a wood charcoal sample embedded within the cement floor of a *chultun* (underground water cistern). This chultun was in service until the Late Classic period, as suggested by a radiocarbon assay of AD 760 ± 100, when refuse above the floor began to accumulate (see table 6.1). The visible buildings of the Grecas Plaza are oriented about 30 to 45 degrees east of north. Substructures beneath the Plaza, the Lintel Building, and associated platforms are between 15–20 degrees east of north and south of east, typical central Mexican orientations of the Early Classic period. These data strongly suggest the presence of a large multi-unit sub-plaza structure constructed in a foreign style similar to those documented at the Platform and Sacta residential groups but located in a central monumental context at Chac.

The Central Acropolis, the monumental core that contains the Grecas Plaza, is the largest and most complex architectural group at Chac (see Smyth et al. 1998:fig. 6). This Acropolis includes a huge basal platform with twelve surface levels faced with slab-shaped boulders covering nearly two full hectares, numerous buildings and features, and three distinct plazas. The Grecas Plaza is centered on the lowest level of the Acropolis and contains a north temple-pyramid (approximately 10 m tall) offset east of north, flanked on east and west sides by two large, double-room temple buildings with I-shaped floor plans.

Approximately 35 m southwest is another temple-pyramid of similar size with a southwestern-facing Megalithic stairway, a smaller western stairway giving access to the plaza, a two-room temple building (Lintel Building), and a stela platform to the south. The plaza is enclosed by two four-room range structures with column doorways; the East Range Structure is attached to a one-room building to the north, while the west structure connects to the double-room building west of the north temple. All structures on the east and west wings are set upon wide building platforms with spacious porticos facing the plaza.

Excavation, consolidation, and restoration took place adjacent to the Grecas Plaza at the Lintel Building (fig. 6.1). The Lintel Building is a two-room Early Puuc–style building on the south edge of the Grecas Plaza named for an enormous in situ interior door lintel measuring approximately 2 m long, 1.5 m wide, and .5 m thick. The doorjambs are made up of well-cut, multiple-piece stone blocks (*sillares*) forming a slightly battered doorway in profile. The west or exterior doorway consists of two large single-piece doorjambs and a large lintel stone. The building shows a two-room I-shaped plan with about 12 m² of floor area for each room. Oddly, the building base, made of well-cut stone blocks, appears to have been laid upon roughly leveled bedrock outcrops. Given the presence of an earlier substructure, it is now certain that these apparent outcrops are really boulder stone walls that were partially collapsed to support the walls of the later superstructure. While the building clearly had a vaulted stone roof formed by triangular-shaped vault stones (but no specialized boot stones like those found on Classic-style buildings), there were no single-member medial moldings so typical as exterior decoration on Early Puuc–style buildings.

An obsidian sample recovered from a 1995 test pit within a ceramic midden just north of the Lintel Building yielded a calibrated obsidian hydration date of AD 667 ± 59, suggesting either a terminal date for the substructure and/or a date of construction for the Lintel Building (Smyth 1998). Exposure excavations for the Lintel Building in 2002 produced more than 12,000 large potsherds associated with this same midden feature. Although mostly Cehpech ceramics, many clearly belong to a defined early phase overlapping with the Motul complex of the Middle Classic period (AD 550–700) (Smyth 1998).

Following excavation and consolidation of the Lintel Building, a 2 × 2 m test pit was placed within the well-preserved stucco floor of each room. The western or outside room revealed the remains of an unusual substructure with two masonry wall segments and well-cut block doorjambs spanning a doorway space of 1.3 m. The substructure likely had

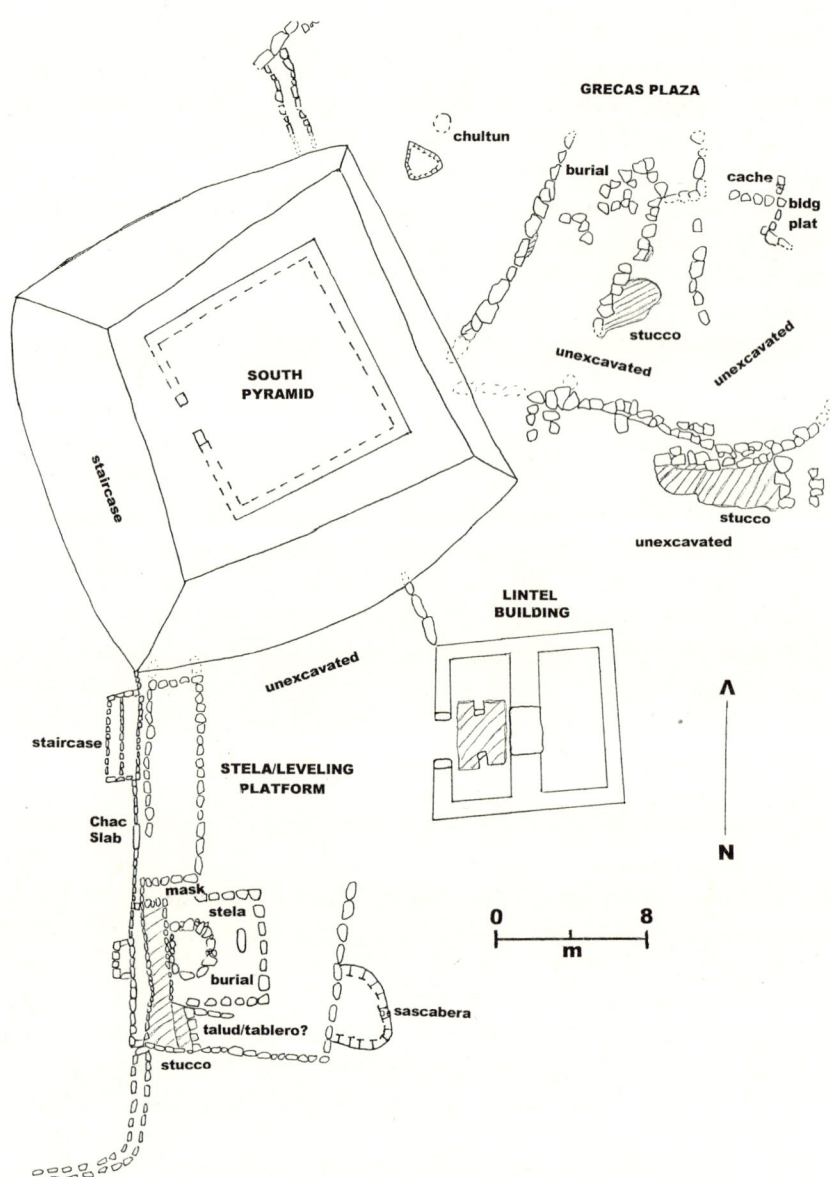

Fig. 6.1 A plan view of the area of excavation at Chac in 2002–2003 showing the Grecas Plaza, the South Pyramid, the Lintel Building, and the Stela/Leveling Platform, and their associated substructures and features. The stucco surfaces represented by hatching indicate stucco floor surfaces associated with the substructure. (Image created by Michael P. Smyth; courtesy of the National Geographic Society.)

multiple-course stone walls with relatively large stone blocks and short tenons, since a stone of this type was found beneath the sealed stucco floor of the interior room and deeply buried near the Lintel Building's northeast corner. All basal and corner stones for the Lintel Building were found intact or nearby.

This substructure, therefore, appears to have had some form of substantial high-walled building, perhaps with a perishable roof, since no early vault stones were found. Although the substructure was partially dismantled upon construction of the Lintel Building, two layers of a distinct reddish-brown stucco floor were found both inside and outside the doorjambs. These floors were laid upon a layer of red earth directly on bedrock similar to a reddish-brown stucco floor surface excavated in 1996 at the Megalithic platform below Building E-VIIa of the Great Pyramid Plaza, radiocarbon dated to AD 520 ± 40 (Smyth 1998). Two other superimposed white stucco floors from E-VIIa directly above yielded wood charcoal specimens and radiocarbon dates of AD 620 ± 50 and AD 700 ± 50 (table 6.1), respectively. These contextual data stratigraphically substantiate the dating of the earliest floor, indicating that the Lintel Building substructure was contemporary with these sixth-century floor terminations.

The Chac Slab

One of the most significant finds of the 2002 season was a sculpted stone, dubbed the Chac Slab, measuring approximately 60 cm square and 20 cm thick. Facing west onto one of the largest plazas at Chac, the slab was set in a retaining wall for a leveling platform for the Lintel Building and nearby stone stela (figs. 6.1 and 6.3). Originally square or slightly oval-shaped at the top, the partially eroded or intentionally defaced upper corners originally exhibited headdresses of two sculpted human figures. Interestingly, the lower left base is completely intact while the lower right corner is clearly broken off. This stone was obviously set in the wall with a broken base, indicating that the slab originally came from somewhere else and must predate the construction of the leveling platform. Architectural stratigraphy shows substructures (discussed below) behind the leveling platform and below the Lintel Building. Because the platform staircase is directly aligned with the Lintel Building doorway and all show similar-style facing stones, they must be contemporary and later in the construction sequence. This means that the Chac Slab must come from another, and perhaps earlier, structure.

Fig. 6.2 The Fine Orange ring-base bowl with tiny supports identified as Provincia Plano-relief. This vessel, found near the foot of the stela within the upper construction fill, shows a scene of a reclining elite figure provisionally identified as a moon deity. The rim diameter is 20 cm and the height is 9 cm. (Photo taken by Michael P. Smyth.)

The bas-relief carving, however, is extraordinary in that it shows two facing elite figures in profile flanking a large exotic bird all about one-third life-size (see fig. 6.3). The left-side figure stands erect. His left arm is adorned with a stone bracelet, and he is holding the arm straight up, while his hand grasps a linear object. The right arm, also with a stone bracelet, is down at the side with the hand holding or cupping the curved end of a stick adorned with knots, jewels, and end feathers. The legs show geometric patterns as if the artist was trying to portray the figure as elaborately clothed. Facing the right figure is a live bird with a long curving neck, crest feathers, and an eye covered by an oval ring or goggle. The bird's left wing is partially open and the right wing is folded in. The feet appear eroded or possibly unfinished, and some sort of protrusion is seen hanging from the breast. The right-side figure shows an unusual facial depiction: a large eye with line markings, an open mouth as if in the act of speaking, and an earflare in the shape of a three-quarter circle. The right arm is extended, holding an offering in the form of a volute or greca symbol in the open palm. A capped, sleeveless jacket or animal skin decorated with flower petals, and perhaps crescent moon signs, is seen below the neck and draped over the left shoulder with hanging elements reaching behind the figure joined by some heraldic-like device. The left arm appears to be down at the side while the left hand is not visible at all. The left leg is slightly bent, with a flower symbol by the knee, and an apparent knot and sack or bundle-like projection is hanging from the rear.

While bas-relief sculpture of the Chac Slab conforms to the Maya narrative style found in the western Maya area during the Late Classic period (Clancy 1985:67–69), the form of the sculpture, certain symbolic motifs, and its context are certainly eclectic and unusual. Carved stone slabs of this size and shape are not common at Puuc sites, which normally contain carved stelae, columns, jambs, and lintels in architectural and plaza contexts. Incorporation into a retaining wall next to a staircase in a prominent public space adjacent to, but not atop, a stela platform is strange indeed. The size, shape, and context of the sculpture are comparable, however, to the Bazan Slab from Monte Albán showing side-by-side Zapotec and Teotihuacan elite figures (Carmona Macias 1993:171).

On the Chac Slab, the left figure appears with some of the emblematic symbols of a Mexican warrior: curved stick (ceremonial atlatl?) and elaborate clothing (Mexican attire?). The curved stick in the right hand and the upraised left hand grasping a linear object(s) is iconographically similar to a Mexican-garbed warrior facing a black-painted lord found on a now-destroyed Early Classic mural from Uaxactún (Martin and Grube 2000:30). The depiction of an exotic bird (perhaps a rare form of currasow, or *k'anbul*) with a goggle eye and open wing is not typical in Puuc iconography. The only other comparable depiction is a miscellaneous sculpture from Labná, possibly a corbel stone from the jamb of a doorway showing birds with linked necks on one side and hieroglyphs on the other (Pollock 1980:51, fig. 101c). The bird figure, however, is also similar to *lechuza y armas* iconography believed to represent Teotihuacan elite military orders (Miller 1973:365; Pasztory 1993:204; Stuart 2000:485–86; von Winning 1985). The eye ring, open wing, and right-facing pose are typical of bird imagery in mural painting at that great highland city.

The right figure, although difficult to discern, appears to portray more typical Maya characteristics of the naturalistic style, including overlapping forms and unearthly associations and symbolism (see fig. 6.3). Although the headdress is completely missing, aspects of the face, hand, and accoutrements, as reconstructed, suggest that this figure may represent a moon deity or priest. The eye and markings, the open mouth, the semi-circular U-shaped earflare, and the open hand and thumb creating an image of a greca could relate to moon signs and conch-shell symbolism. Schele and Miller (1986:308–9) have interpreted and deciphered a scene and inscription on a carved Early Classic conch shell as representing the Moon Lord and God Y, the god of the conch shell who announces the arrival of the Vision Serpent. Interestingly, the right figure of the Chac Slab appears to display a protruding element from the chin, which may be a serpent beard. Another potentially important identifying feature is the garment seen around the neck hanging down and back over the left shoulder. This garment may be a *quechquemitl*, a sleeveless short cape that the Maya adopted from Teotihuacan in the Early Classic period (Taylor 1983:1:72). Normally worn by elite Maya women, the flower motifs, linear markings, and grid patterns seen on the figure suggest that the garment was embroidered or perhaps overlain with a beaded pattern. Beaded garments are closely associated with the moon goddess and her supernatural family such the Hero Twins (Kerr 1992:65; Taylor 1992:517). The flower motifs are of particular interest because they extend from the shoulder area down below the knee, much the way a flowering vine might be symbolically portrayed. In fact, there is a plant with medicinal properties in Yucatán called *zutup* (spp. *Impomoea bona-nox*), or the moonflower vine, that has large white flowers that open late in the evening (Standley 1920–1926:1201).

About half of a Fine Orange bowl with a pedestal base and tiny knobbed supports (the Chac vessel) was recovered within the upper construction fill for the stela. Identified as Provincia Plano-relief, this pedestal base bowl appears to date to the Late Classic period but could be even earlier (fig. 6.2). A white interior and exterior slip with plano-relief and incised decoration, the Chac vessel shows two double-line panels (originally four) with another double-line border just below the rim and repeating slab-shaped forms with small central circular designs. This important vessel shows iconography similar to the Chac Slab, especially the right figure, and is strikingly similar to one known to have come from Moxviquil, Chiapas near Palenque (R. Smith 1971: 86–87, fig. 58a). Most important is an extraordinary scene of a reclining male figure looking left showing unusual facial characteristics. While the nose and combed-back hair seem foreign, the crescent markings above and forming the eye and the linear marks above and below the open mouth, as well as the three-quarter circular earflare, suggest moon deity signs, much like the right figure in the Chac Slab. The necklace, bracelets, and earflare demonstrate the elite or unearthly status

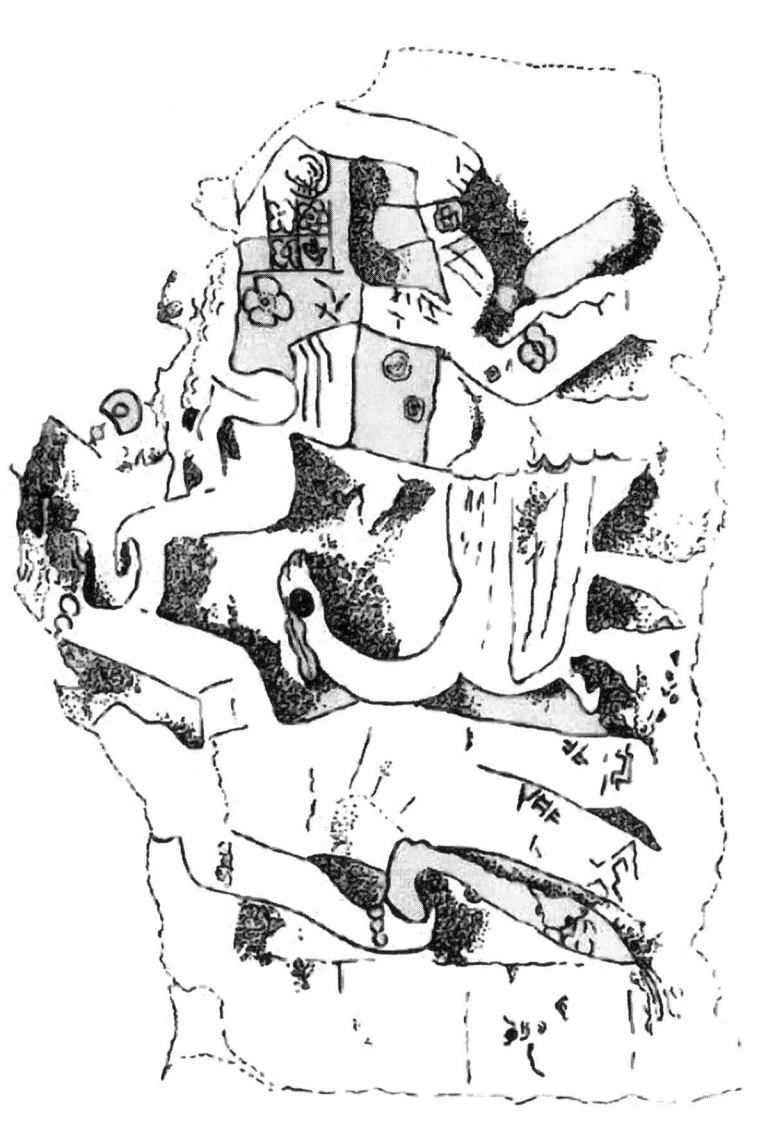

Fig. 6.3 The Chac Slab recovered within the outer retaining wall of the Leveling/Stela Platform. Exhibiting a broken base on one side, this stone sculpture came from an early architectural context. The scene suggests an arrival showing two human figures surrounding an exotic bird (currasow or *Kanbul*). The left figure may be an arriving foreign elite warrior and the right figure is provisionally identified as a moon deity or priest. (Drawing is by Shane Gray; used with permission.)

of the figure, and the crescent attached to his back is closely identified with lunar deities (Kerr 1989:20; Schele and Miller 1986:plate 121; Taylor 1992:519). Open palms on left and right hands invoke a greca or conch shell that may be associated with Vision Serpent symbolism. The conch motif is surrounded by cursive *M* signs, which may stand for the word *U* or moon (Schele and Miller 1986:309). Cursive *M* signs are also seen on the right side of the panel of a crescent-like motif and may even form the toes of the right foot.

Although preliminary, the scene on the Chac Slab can be interpreted as commemorating the arrival of an important foreigner who is greeted by a moon deity or priest. Whether the symbolism depicts a mythological encounter or is a metaphor for an actual arrival is difficult to determine. The body gesture of the left figure holding a ceremonial spear launcher on the curved end in a non-aggressive, non-lethal position does imply peaceful intentions. The context of the slab located on the west side of the platform facing a large plaza suggests that the arrival either emanated from, or was symbolically affiliated with, the west, the direction of the setting sun as well as of the great highland metropolis. The presence of an earlier construction behind the slab stone, its broken base, and eroded/defaced condition indicate that this stone came from an earlier structure. Whether the slab was contemporary with a supposed arrival event and reset in the retaining wall of the leveling platform at a later time is difficult to reconcile with the current data. The one-third life-size of the figures also suggests that the stone originally came from another architectural context not associated with the public plaza. A life-size or larger-scale representation that can easily be seen during large public gatherings is expected in this context. Other possibilities are that the arrival took place at an earlier time and is being memorialized, or even that the alleged meeting between the foreigner and Maya deity was mythological, occurring in the spirit world. Other data from the Grecas Plaza during the 2003 field season (see below), however, suggest an actual presence of elites and their retainers with affiliations or even origins outside the Maya area.

An Early Substructure

The 2003 work now provides evidence for substructures beneath the Leveling/Stela Platform, the Lintel Building, and south half of the Grecas Plaza itself (see fig. 6.1). There are at least three construction phases associated with this complex architectural space. Excavations of the Leveling Platform revealed an earlier wall alignment with rough-shaped boulders set upon leveled bedrock behind the faced-stone retaining wall, the staircase, and the Chac Slab. This boulder wall continues underneath the South Pyramid in a north-northeast direction.

Recovered on the south side of this wall within an almost certainly interior surface was a great quantity of stucco fragments in a variety of modeled forms and painted colors. These include malachite green and turquoise blue beads of a heavy necklace, geometric border elements suggesting a panel of sorts, incised grecas, headdress feathers, a red earflare, and incised rows of stucco panels decorated with columns of small cream-colored ovals (perhaps representing *cacao* beans) set upon a red field. There were also many miscellaneous pieces painted other shades of red, blue, black, ochre, and red specular hematite. Likely destroyed during a later phase of construction, these remains indicate that there was a stucco mask-panel representing a frontal portrait of an elite human figure painted in a spectrum of polychrome colors.

Running to the south is a strange winding masonry wall of irregularly shaped facing stones attached to the boulder wall for the stucco mask. This masonry wall ranges from 80 cm to almost 2 m wide, is about 1 m tall, and shows a stucco layer along most of its superior surface. Widening to the south, the stucco surface slopes sharply upward to become integrated into the base of an upright boulder wall, forming a slope and panel façade near the substructure's southwest corner.

The stucco covering of the upright boulder wall has long ago disintegrated, but the ratio of boulder height (*tablero*?) to the sloping stucco surface height (*talud*?) is approximately 3:1, the typical talud-tablero relationship found at Teotihuacan. A strange, round tenon stone carved with a circular incised border motif (eye or shield?) was found at the bottom of the substructure fill within an apparent room area adjacent to the slope and panel facade. A heavy boulder wall running in a direction of 15 degrees south of east appears to define a room area, then turns north about 5 m east and follows an east of north heading towards the Lintel Building substructure. It is argued that these wall alignments defined the south portion of a multiple-room, modular-style building once integrated with substructure remains to the north.

A circular stone-lined cist about 2 m deep and 1.5 m wide was constructed within the interior area of the substructure after it was dismantled, collapsed, and filled in. Buried deep within construction fill, the cist contained ash and numerous remains of burned human bone —representing two and possibly three cremated burials—but no complete or significant artifact offerings. Located midway between the stone stela to the east and a stone step to the west, the cist feature appears to represent a cremation offering and termination event for the substructure occurring before later remodeling. At some point follow-

ing the termination and filling of the substructure *and* after the placement of the cremated burials, a square platform and a plain-stone stela
standing 1.8 m tall were erected over the substructure about 2 m east of
the cist. It was at the foot of this stela that the Fine Orange Plano-relief
vessel mentioned above was recovered. About 2 m to the south, a large
tenoned stone decorated with two dots in low relief on each side was
found in the upper levels of stone fill. This stone appears to be a projecting device or cord holder carved to imitate a human foot or sandal
occasionally found on vaulted Puuc buildings (Pollock 1980:574–75).
It is not known from where this stone came or why it was removed and
added to the fill here, but it must have been near the end of occupation at Chac.

A stone retaining wall of faced stonework, including the Chac Slab,
completely encapsulated the substructure, and a two-step staircase was
constructed to the north to give plaza access to the Lintel Building.
A *sascabera* (limestone quarry) located under the east wall of the substructure appears to have been used primarily as a quarry for construction material for the Lintel Building. The final phase of construction
was a double stone wall made up of reused stones from nearby buildings constructed to close off access to the Grecas Plaza area.

Horizontal excavations in 2003 opened the south half of the Grecas
Plaza north of the Lintel Building between the South Pyramid and
East Range Structure (see fig. 6.1). Various stone alignments in this area
are at a higher surface level but relate spatially to the substructures
beneath the Lintel Building and Leveling Platform and show that an
earlier building is beneath the level of the plaza floor. Stucco floor fragments for the late plaza surface were found superimposed over a series
of boulder walls placed upon leveled bedrock and later collapsed and
filled in. Although partially destroyed, there are two wall alignments
of large boulders that intersect somewhere beneath the South Pyramid. Estimating their approximate point of intersection, the west wall
runs about twenty-five degrees east of north, and the south wall runs
about fifteen degrees north of east. A series of parallel and traverse
stone alignments were found, defining possible room divisions with
wall and floor stucco in situ, clearly showing that this was an interior
space of a multi-unit building and not a leveling platform.

A seated burial within a circular stone-lined cist was encountered
along with three early-style ceramic vessels: (1) a Kinich Naranja hemispherical bowl; (2) a Say Slateware hemispherical bowl; and (3) a Say
Slateware tripod dish with nubbin supports and thumb-impressed
decoration. All of these vessels belong to the Motul complex at Chac
and date to the Early Classic–Middle Classic period. In another partial-

cist feature to the south below the substructure floor level, fragments
of human bone and a complete Chemax slateware burial bottle (venenera) similar to others found in burials at the Platform and Sacta residential compounds were found (Smyth and Rogart 2004). Burial bot-
tles of this type are extremely rare in the Puuc region but date from the
sixth to seventh centuries at Chac.

A major substructure wall curves towards the northeast and passes
underneath the East Range Structure. Perpendicular wall segments to
the south clearly continue well beyond the horizontal exposures and
define room areas or corridors surrounding a patio area to the west.
Well-preserved stucco floor surfaces were found at the bases of these
stone alignments and strongly indicate that the substructure continues to the southeast and northeast. A probing excavation below the
southwest corner of the building platform for the East Range Structure revealed another substructure wall running to the west beneath a
large striated vessel containing multiple offerings that was clearly associated with the later building platform. Another probe to the northwest encountered two rows of facing stones defining part of a staircase
descending into the Grecas Plaza, as well as a boulder wall that appears
to be part of a substructure passing below the building platform for the
West Range Structure. A partially collapsed chultun lies between the
boulder walls.

This amazingly complex architectural space appears to have been
constructed and occupied from the Early Classic to the Late Classic
periods, undergoing several episodes of filling in, rebuilding, and
modification resulting in the Grecas Plaza, the Lintel Building, and
the Leveling/Stela Platform. The Phase I substructure, however, differs
significantly from the later construction phases because it appears to
be a large composite building. Although some architectural details are
not clear because the substructure was partially dismantled, its spatial
organization viewed at a larger scale suggests that there were room-
block areas, corridors, and interior patios oriented in a general north-
to-northeast direction. Multiple surface levels are also found that are
typical of apartment compounds at Teotihuacan: the highest elevations are at the Grecas Plaza sub, followed by the Lintel Building sub,
and then the Leveling/Stela Platform sub. Although the significance
is unclear, two kinds of early-style stone masonry are integrated into
the substructure: rough boulders and small irregularly shaped facing
stones, apparently mixing Megalithic and early Oxkintok styles.

Although the ceramic analysis of the 2003 season is still in progress,
general trends suggest that most of the ceramics from the substructure are striated and slateware sherds (including Muna and Say) with-

out decoration, including early resist-painted slatewares like Chemax of the Motul complex. This pattern is not typical of the Cehpech ceramic assemblage from later monumental contexts at Chac. The normal Cehpech wares of Red Teabo and especially Thin Slate wares are comparatively rare, but an unidentified Thin Orange (mostly of Kinich Naranja group) is consistently found at the lowest levels, often mixed with a scattering of Early Classic polychrome and slateware sherds. These same ceramic patterns are found at the lowest levels of the Platform and Sacta groups residential compounds (Smyth and Rogart 2004) and clearly show that Motul complex ceramics, including early slateware pottery, date to AD 500, if not earlier.

Lithic patterns are also quite suggestive. The relatively high frequencies of obsidian associated with the Grecas substructure ($n = 32$) is comparable to frequencies at the Platform and Sacta groups, where more than half of all the obsidian found at Chac before 2003 had been recovered ($n = 92$). This is not a product of sampling because more architectural and stratigraphic excavation by volume has taken place outside of these particular substructure contexts. While instrumental elemental sourcing of all the obsidian from recent field seasons is pending, visual identifications aided by a sample of obsidian specimens sourced by neutron activation in 1996 (Smyth et al. 1998) estimate that more than 10 percent of all obsidian at Chac can be attributed to non-Guatemalan sources. Even though there is little obsidian from substructure contexts that can be definitely traced to the Pachuca source (Pachuca obsidian has been found in both early and late architectural contexts at the Great Pyramid Plaza), there are a number of other flakes, blades, and biface fragments that appear to be Otumba, Zaragoza, or some other Mexican obsidian.

In addition, the most common biface tool form from these substructures is a small chert point identified as a dart point for use with a spear launcher (atlatl), a weapon emblematic of a Central Mexican warrior. There are also two obsidian biface points, one from an early Pyramid Plaza offering and the other from the Sacta group substructure that show typical Teotihuacan lithic workmanship (Spence personal communication, 2003)—a third reworked biface from the Grecas Plaza sub does as well, and is likely from Mexican sources.

Discussion and Conclusions

The role that Teotihuacan played in the Maya lowlands continues to be a subject of intense debate and controversy. While Maya centers such as Tikal, Kaminaljuyu, Balberta, and Copán may have experienced some form of direct "takeover," other sites like Altun Ha and those to the north such as Becán and Oxkintok show less direct Teotihuacan influence. Large contemporary sites such as Calakmul and Edzná show little or no evidence of Teotihuacan influence at all. Such wide-ranging positions may be partly a result of limited sampling but also certainly relate to what Marcus (2003:344) refers to as "cyclical interpretation," where new generations of Mayanists return to previous positions or resurrect old, discarded ideas. Moreover, great variability of interaction must relate to the fact that Teotihuacan had different relationships with different Maya centers at different times. This complex cultural picture cannot be readily understood by "either/or" positions, particularly when it has become increasingly evident that there were many other culture regions including the Gulf Coast, Oaxaca, and the Maya highlands that played a critical intermediary role in relations between Teotihuacan and the Maya lowlands (Marcus 2003:342). And what can be said of the Puuc region? Were early Maya centers here outside the sphere of influence of Teotihuacan and other Mesoamerican centers?

The presence of Mexican decorative elements and symbolism on Puuc architecture has long defied explanation. Research at the Maya center of Chac now suggests that early Mexican symbolism is indeed related to a foreign elite presence emanating from Teotihuacan and its surrogates, beginning near the end of the Early Classic period (AD 300–550). Intensive excavation at two residential groups at Chac support the identification of foreign-style residential architecture, artifacts, and mortuary patterns similar to Teotihuacan apartment compounds dating from the Middle Classic period (AD 550–650) or earlier. The fact that these residential compounds lie outside the site's monumental core and contain low quantities of elite materials indicates that the occupants were not of high status; perhaps they were foreign merchants or trade representatives from the highland metropolis who married into local Maya populations (Smyth and Rogart 2004).

Artifacts and early architecture at the Great Pyramid Plaza, however, incorporate icons and symbolism associated with the Teotihuacan style. In addition, the research at the Grecas Plaza in 2002 and 2003 shows unusual elite iconography, the remains of a substantial non-Puuc substructure, Thin Orange–like pottery, and slope and panel decoration. These data suggest that foreign relationships were at a high social level and went beyond long-distance influence or simple emulation by local elites but perhaps not as far as outright political or military domination. There appears to be some other complex form of mid-range interaction in play at Chac.

The provisional data suggest that there were foreigners at Chac of elite status who resided in the Grecas Plaza. Sub-plaza contexts of the Grecas Plaza clearly show a large early substructure with stonework, spatial conventions, and monumental context indicative of a palace-type building in a foreign style. The stucco remains of a talud-tablero-like façade showing the typical 3:1 ratio (panel height to sloping wall height) at Teotihuacan and remains of a polychrome stucco mask-panel from an interior context were found. Similar painted stucco remains were found associated with the Middle Classic phase of the Great Pyramid constructed and decorated in a foreign style. A stone-lined cist containing multiple cremated burials, but no interments, placed within a substructure interior space (room?) in front of the stela, but at a much lower surface level, suggests a sacrificial offering for a termination event associated with the substructure. Mortuary patterns of seated, flexed, and now cremated burials placed within circular stone-lined cists are typical of Chac and closely comparable to round burials commonly found at Teotihuacan.

Whether foreign elites were actually central Mexicans, affiliated with central Mexico but from some other Maya center, or local Maya who had lived in Teotihuacan is difficult to determine with the available data (see Taube 2003). A strontium analysis of human skeletal remains is underway and, hopefully, should help to resolve some of these ambiguities. It seems more probable, however, that long-distance contacts with Teotihuacan were maintained via intermediate centers such as Becán in Campeche, Matacapan in Veracruz, and Tikal in the Petén.

Ceramics from Chac, for example, show similarities to wares at Becán and Middle Classic Matacapan. Early vessels with forms and decoration in Teotihuacan style at Chac are identified as non-local wares with likely origins from somewhere on the Gulf Coast. Orange-ware vessels and sherds recovered from substructures do not appear to be Thin Orange wares from Tepexi de Rodríquez, Puebla, nor San Martín Orange from Teotihuacan. Significant quantities of orange-ware sherds and one Thin Orange bowl (Dzilam Naranja Acanalado) recovered from a substructure context at Chac, however, do show surface treatment similar to large deep bowls, or craters, of San Martín dated to the Late Xolalpan phase (AD 450–550) including "rounded bases with pocked or roughened surfaces and the remainder of the exterior has a distinctive finish, i.e., light striations suggestive of scraping with a serrated instrument" (Rattray 2001:237).

Probably used for cooking and/or serving at Chac, the Thin Orange bowl shows evidence of having been made using a mold and coils similar to that described for the production of cooking pots at Teotihuacan (Rattray 2001:237). Dean Arnold has suggested, based on cross-cultural pottery ethnographic work, that forming technologies are one of the most conservative of all ceramic attributes and may be an indicator of ethnicity and the transport of the potters themselves (D. Arnold 2003:507). Were potters at Chac familiar with Teotihuacan ceramic making or were foreign potters present at Chac?

Lithic data point to the presence of El Chayal obsidian from highland Guatemala as the dominant source. However, Mexican (Pachuca, Otumba, and Zaragoza) obsidian is also present. The source data imply that Teotihuacan was involved in the movement of obsidian across the Maya area. As suggested above, the thin obsidian bifaces from Chac may have been manufactured in Teotihuacan. Did they arrive as down-the-line trade, as gifts, or with visiting foreigners? It is interesting to note that Moholy-Nagy (1999:302–4) suggests the possibility that all Early Classic–Middle Classic thin bifaces of green and gray Mexican geologic sources in the Maya area, especially at Tikal, were manufactured at Teotihuacan. It may have been the quality of workmanship, interestingly, that was in such high demand among the Maya and not the exotic color or quality of the obsidian itself.

The important new data from Chac show that the internationalization of the Yucatán began long before the Terminal Classic period and place the Puuc region within the sphere of influence of the greatest polity of Classic Mesoamerica. Why would Teotihuacan be interested in such a far-away place as Chac at a time approaching the end of the great highland city? We have argued that Chac was located along a strategic overland trade route connecting the Puuc region to the northern coastal plains, south to the Maya lowlands and highlands, and west to the Gulf Coast and central Mexico. The nearby Gruta de Chac, or Chac Cave, is part of the Chac site (Smyth 1999; Smyth and Rogart 2004) and contains the only permanent water source for miles around. As a sacred pilgrimage site in the Early Classic period, the cave may have held particular fascination for Teotihuacanos. While caves were sacred to the Maya and all Mesoamerican peoples, they were of special significance to Teotihuacan because the cave beneath the Pyramid of the Sun was seen as a place of origin, where the moon was born, and where time itself began (Carrasco 2000:108; Heyden 1975:139).

Although hypothetical and subject to considerable future research, Chac's attraction to Teotihuacanos and others may have included its value as a rich source of medicinal products and spiritual-healing knowledge of Maya priests. While there were significant centers in the region during the Early Classic period, the Puuc region was not heavily populated or deforested until the Terminal Classic period. This means

that the region could have produced an abundance of important resources including honey, wax, tobacco, and a cornucopia of powerful
medicinal plants. As trader bundles, these kinds of items could have
been easily and effectively traded over great distances; there was a brisk
trade in many of these items in the Yucatán during the Late Postclassic
period (Roys 1943, 1957).

By the sixth century, Teotihuacan was suffering from disease and
sickness related to poor sanitary conditions brought on by overcrowd-
ing, hyper-urbanism, and perhaps extended drought (Story 1985,
1992); medicinal remedies and medical treatment must have been in
great demand and of paramount importance to Teotihuacan's elite
class. At about the same time, political instability and endemic warfare that engulfed many southern Maya cities must have significantly
reduced or severed altogether Teotihuacan's trade routes in the southern lowlands. Such a scenario would have required opening new trade
routes or intensifying old ones to places like the northern Yucatán
where alternate supplies of forest products and other trade goods could
be more easily obtained. Arriving Teotihuacanos seeking medicinal
remedies and spiritual cures are not outside the realm of possibilities,
and actual visitations are suggested by the data at Chac. Such a long-
distance movement of people was most certainly not one-way, as Maya
traders and/or elite would have been eager to visit the most holy site
and greatest pilgrimage city of the Americas. Perhaps this is why there
are so many Maya ceramics from the northern lowlands at the Merchants Barrio of Teotihuacan (Rattray 1987:267).

The dynamics of early foreign contacts, directionality, duration, and
intensity are important issues that have not been comprehensively explored in the Puuc region and northern Yucatán. Ball's (1994:394–95)
call to arms a decade ago to address key questions regarding the Puuc
origins still resonates resoundingly today, even though slow progress
has been made in recent years. For example, the existence of significant centers in the region during the Early Classic period has been
confirmed, at least for sites such as Oxkintok and Chac (Varela Torrecilla and Braswell 2003). It seems reasonable to ask, therefore, whether
there are other sites in the region as early or earlier; the answers, unfortunately, often lie deeply buried beneath Late Classic and Terminal
Classic construction. The role of long-distance trade and economic exchange in the Puuc urbanization process and the relative importance
of foreign interactions will remain unanswered questions until considerable research attention becomes directed towards resolving these
fundamental problems.

Nearly a decade of research at Chac is beginning to reveal a previously unknown chapter of northern lowlands prehistory. The origins of the Puuc cities and the nature of the region's early political economy are being addressed by the work at Chac. In addition, evidence can be marshaled for a foreign presence, including political elites, at the end of the Early Classic period. Although these findings were certainly unexpected, they become more plausible given the early foreign influence documented at nearby centers like Oxkintok. Indeed, current data support the idea that Chac maintained far-reaching foreign contacts with the southern Maya area, the Gulf Coast, and central Mexico, including the powerful highland metropolis of Teotihuacan. Such data will continue to transform ideas about the nature of complex societies in the Puuc region during the Early Classic period and the significant roles they played in the great Terminal Classic florescence of the northern Yucatán.

7

Classic Politics in the Northern Maya Lowlands

Justine Shaw and Dave Johnstone

"Politics" is defined as a system of governance within and between political entities such as sites, alliances, and states. Archaeologically, political systems leave traces in the written record (where present), the movement of goods between centers, *sacbe* (road) systems, architectural styles, settlement patterns, and the warfare events detected at many sites.

Using hieroglyphic texts, Mayanists are reconstructing the political structure of the ancient Maya. For the majority of the Classic period in the northern lowlands, polities were organized according to the principle of divine kingship (Schele and Freidel 1990). Kings, or *ahauob*, served as religious leaders and civic heads of state at sites throughout the Maya area. *Kalomte*, the title for a war leader, appears to be an office superior to *ahau* (king), as these individuals ruled more territory than that of a single site (Harrison 1999; Stuart et al. 1989). The title *sahal* appears to have referred to subordinate officials that ruled towns for their overlords and served as war captains and court officials (Schele and Mathews 1998:89).

These titles imply a hierarchical arrangement of sites, but as yet, the size and composition of political units is a matter of debate. Mayanists at one end of the spectrum view each site as a totally independent political entity (P. Mathews 1991) in which all ahauob were of equal status with political ranking only existing within the city-state. Alternately, others (Adams 1986; Marcus 1973) envision regional states based upon the distribution of emblem glyphs. This hypothesis includes regional sites that were more powerful controlling lesser centers. Martin and Grube (1995) propose a larger alliance system based upon two competing superpowers in the central lowlands—Tikal and Calakmul.

Recent research on the topic of Classic Maya politics in the northern lowlands has primarily focused on Chichén Itzá (A. Andrews 1990a; A. Andrews and Robles Castellanos 1985; Krochock 1998; Schele and Freidel 1990), due in part to the well-preserved written record from the site. However, the limited temporal and geographic data upon which these models are built make it difficult to apply them directly to other northern sites.

In comparison to the southern lowlands, the northern lowlands have relatively few readable hieroglyphic texts. Part of the paucity of readable texts is due to the poor geological quality of the limestone available in the region. Texts carved on this soft stone rapidly weather when exposed, leaving many stelae indecipherable. However, based upon readable monuments, some statements can be made about the nature of northern politics. While the northern site with the greatest number of hieroglyphic texts, Cobá, has yet to produce an identifiable emblem glyph, such glyphs are known for the sites of Dzibilchaltún (Schele et al. 1998), Chichén Itzá (Grube 1994), Ek Balam (Vargas de la Peña and Castillo Borges 1999), and Uxmal (Kowalski 1985, 1987).

The emblem glyphs of Uxmal and Ek Balam record the *k'ul ahau* title as a part of their place names, suggesting that divine lords led these polities. Glyph blocks from a dismantled hieroglyphic stairway at Yo'okop include the title "kalomte" and the phrase *u kahi*, translated as "by his doing" or "under the auspices of" (Martin and Grube 1995). This suggests a hierarchy within the class of ahauob, with some being politically subordinate. Another title, "sahal," is found at the site of Mopila (Freidel 1992), suggesting that it was a secondary site to a larger center, probably Yaxuná.

The presence of political titles and the distribution of emblem glyphs suggest that some type of hierarchical political system headed by divine kings was common in the north. However, it was not the only option, as Xcalumkin and Chichén Itzá experimented with joint rule, or *multepal*, of lords with the same status (Krochock 1998; Schele and Freidel 1990).

Additionally, we can infer northern political relationships based upon textual references to other sites and events. At Yo'okop, Early Classic glyph blocks refer to Calakmul's Ruler 17, "Sky Witness" (Martin 1997; Shaw et al. 2000). Cobá records the arrival of a princess from Dos Pilas at Naranjo in AD 682 (Schele and Mathews 1998:202). Schele and Mathews (1998) have suggested that Cobá was a member of an alliance system led by Calakmul.

Since most northern sites lack well-preserved hieroglyphic texts, researchers have had to find other ways to detect political entities in the archaeological record. Therefore, it is necessary to devise a means to utilize the existing ceramic and architectural evidence to make inferences about the nature of northern Maya macro-political systems.

The ceramic assemblage of any site is the result of a number of processes, including local production, trade, tribute, plunder, and gift exchange. As such, the ceramic complex (the total ceramic content of a site during a period) is the product of cultural, economic, political, and historical processes. Owing to these factors, we would not expect that any two sites would have identical complexes.

Similarities between ceramic complexes do exist. Those complexes sharing a majority of their most common types are considered to belong to a common ceramic sphere. As ceramic vessels are portable artifacts, individual types are not sphere-specific. These spheres imply a common ceramic tradition and a high degree of technological contact between member sites and are the basis from which other cultural units might be inferred. For example, the replacement of one ceramic complex with another one containing intrusive elements has been interpreted as representing intrusion of foreign peoples, such as with the Floral Park complex (Willey et al. 1967). The distribution of related complexes belonging to the Floral Park sphere then might approximate the extent of foreign displacement.

Traditionally, this process has been an intuitive effort based on the experience of the individual ceramic analyst. This can be problematic, resulting in competing classifications. For example, Late Classic Cobá has been included in the Motul (Smith and Gifford 1965), Copo (Ball 1978), and Tepeu (Robles Castellanos 1990) ceramic spheres. In an attempt to measure the strength of association between ceramic assemblages, the correlation coefficient was adopted by Johnstone (2001) as a tool to assign a numeric value to the degree of similarity between the ceramic complexes of different sites. Sites with R-values greater than 0.6 exhibited a strong degree of association and were considered members of the same ceramic sphere.

Unfortunately, the application of ceramic data to temporal questions has overshadowed their application to cultural reconstruction. The "linear succession" model (Brainerd 1958; R. Smith 1971) grouped ceramic complexes into pan-peninsular entities corresponding to temporal horizons. A comparison of ceramic assemblages from sites across the northern lowlands (Suhler et al. 1998:fig. 4) shows little correspondence in the duration of phases or in their temporal placement. This suggests that, rather than broad regional stages or horizons in which new traditions evolved and gradually replaced older traditions, the timing and composition of local ceramic complexes may be due to specific historical events.

More recently, the "total overlap" model (Ball 1979) has been proposed in which multiple ceramic spheres contemporaneously coexist. The acceptance of the total overlap model presents a number of difficulties, such as how to account for radically different ceramic assemblages at sites located short distances from each other, and how to account for replacement of one complex with another containing new types or traditions.

Wholesale change in a ceramic inventory requires that the producers adopt new design and production modes or that there be a replacement of local ceramic specialists by specialists familiar with a different ceramic tradition. For a change in rulership to have a significant effect on the production of ceramic types and forms, production should be either highly centralized or under the direct patronage of the incoming elite. Many sites, such as Tikal, had multiple centers of utilitarian ceramic production (Fry 1979). Some sites had "palace workshops" responsible for making fine ceramics for local consumption (Ball 1993). In the case of utilitarian ceramics with decentralized production, we would not expect elite patronage to play a significant role in determining the nature of the end product or of its distribution. However, fine-paste ceramics represent centralized production for a restricted market in which elite patronage has a more significant role in both the nature of the product and in its distribution.

Specialized ceramic wares were traded greater distances than utilitarian wares, suggesting a different mechanism for their distribution, with elite gift exchange a likely possibility. Gift exchange, including fine ceramics, was an important means of cementing political alliances and larger-scale trading networks (Ball 1993; Reents-Budet 1998).

If trade was free and open, then the only impediment to the distribution of ceramics should be distance. For much of the Maya lowlands, non-elite ceramics had a restricted distribution of 30–50 km (Fry 1980:10). The wide distribution of utilitarian ceramics cannot be explained by market forces and are likely the result of shared tradition. Differences between spheres are the result of having different production traditions and a lack of significant trade between the regions. One possible explanation for this trade barrier is a political one, in which a state of hostilities existed between the regions.

When foreign ceramics that may be attributed to a given site are found in contexts of warfare and destruction, it may be hypothesized that the bearer of the foreign ceramics was responsible for the event. Adams (1999) notes that Río Azul, a site on Tikal's northern border allied by marriage and descent to Tikal, also experienced conquest associated with a change in ceramic complex. The Teotihuacan-inspired pottery at Río Azul was smashed into small pieces and replaced by a ceramic complex containing all new types.

Architecture may also serve as a means to explore politics in the absence of a written record. Specific architectural styles may be said to be diagnostic of certain localities, although styles generally reflect regional practices, rather than single sites (see Mathews and Maldonado Cárdenas, this vol.). Since architecture is not portable and takes longer to create, the architectural style(s) present at sites may indicate a more permanent influence at a site. At Dzibilchaltún (E. Andrews IV and Andrews V 1980), architectural change accompanies what appears to be a political transition at the beginning of the Terminal Classic period. "Transitional vaulted" buildings represent local masons experimenting with new techniques based on their Late Classic (Copo I phase) experience (E. Andrews V 1979; E. Andrews IV and Andrews V 1980). This syncretism results in a number of hybrid buildings combining load-bearing and veneer elements. Construction is often executed onto, or over, existing early facet Copo I architecture, with an effort made to carefully bury the earlier structure. In these cases, the Terminal Classic (Copo II) masons remained faithful to the original layout of the site, replacing older buildings with larger, more elaborate editions.

A final means to examine past politics is through *sacbeob*. Like other forms of architecture, the enormous investment needed to create these roadways means the link between any two points connected are of a relatively long-lasting nature. Archaeologists can examine the sites/points connected by intersite roadway systems to give insights into political connections. These data may be combined with artifactual evidence, such as looking for diagnostic ceramics associated with road construction. Cobá's sacbe system provides a clear picture of the areas firmly under the site's political control, with forty-five sacbeob extending from 6 m to 100 km from points within the site core (Benavides Castillo 1975). Caracol (D. Chase and Chase 1992) and Calakmul (Folan 1992) similarly seemed to use roadways to define and manage areas under the political control of these sites (Shaw 2001).

Current Research

While ceramics, architecture, and sacbe systems have been used to define "culture areas" or regional interaction zones, they have seldom been used in concert to attempt to identify the participants in specific historical events and processes. At Yaxuná, researchers have used such a "conjunctive approach" at a site where glyphic texts are largely absent. When David Freidel began research at Yaxuná, he stated two hypotheses. The first was that the 100 km Sacbe 1 was built in the Terminal Classic period by Cobá as a response to Chichén Itzá's aggression in the region. Investigators hypothesized that Cobá occupied Yaxuná in the Terminal Classic period, at a time of peninsula-wide struggle with Chichén Itzá for control of the northern Maya lowlands (A. Andrews and Robles Castellanos 1985:66; Freidel 1992:39). The second hypothesis stated that Chichén Itzá and Yaxuná were in direct conflict, based on ethnohistorical accounts in the *Chilam Balam* (Roys 1933).

The Yaxuná Project has used a variety of data to examine political interactions. When ceramic data were integrated with architecture, burial, sacbeob, and settlement data, a more detailed sequence could be constructed and placed in its regional context. Using multiple categories of data gave researchers insights that were more detailed and gave them greater confidence in their interpretations.

During the Early Classic period, Yaxuná provided evidence for divine kings and cultural ties with the northwest portion of the peninsula. One royal burial was excavated dating to this period. The king was in primary context, placed in a vaulted tomb with a large quantity of elite grave goods (fig. 7.1). Several important public buildings were constructed in the Megalithic style at Yaxuná (see Mathews and Maldonado Cárdenas, this vol.). Ceramically, the Yaxuna IIa phase appears to have been affiliated with the Xculul ceramic sphere characteristic of sites in the northwest portion of the peninsula (Johnstone 2001). Polychrome trade wares from the Petén were relatively abundant during this period, suggesting strong trading links between these regions. Although Yaxuná cannot be tied to a specific site at this time, its political associations are thought to follow its ceramic affiliations.

The Middle Classic period began with a complete disjunction in the ceramics and architecture of the site, accompanied by a change in the dynastic succession. Architecturally, Yaxuna IIb saw the introduction of apron moldings with inset panels. The arch became a simple cantilevered vault, replacing the stepped vault. Thick, modeled stucco decoration was also introduced. These architectural traits are characteristic of Oxkintok (Rivera Dorado 1991) and were incorporated into the construction of Structure 6F-4–3rd, a construction phase built to cover the tomb of Burial 24. This tomb contained the remains of eleven elite men, women, and children. These remains have been interpreted (Freidel and Suhler 1998; Suhler and Freidel 1998) as the sacrificed remains of Yaxuná's rulers, including the decapitated king and the rest of his family. The ceramics cached in Structure 6F-4–4th, prior to its being covered by Structure 6F-4–3rd, were all types common to Oxkintok. One of these contained a set of royal crown jewels that were ritually "killed" (Suhler and Freidel 1998:33) by an axe left in the vessel. Yaxuná's Stela 1, probably originally set in front of Structure 6F-4, depicts a ruler attired in Teotihuacanoid garb associated with Venus-

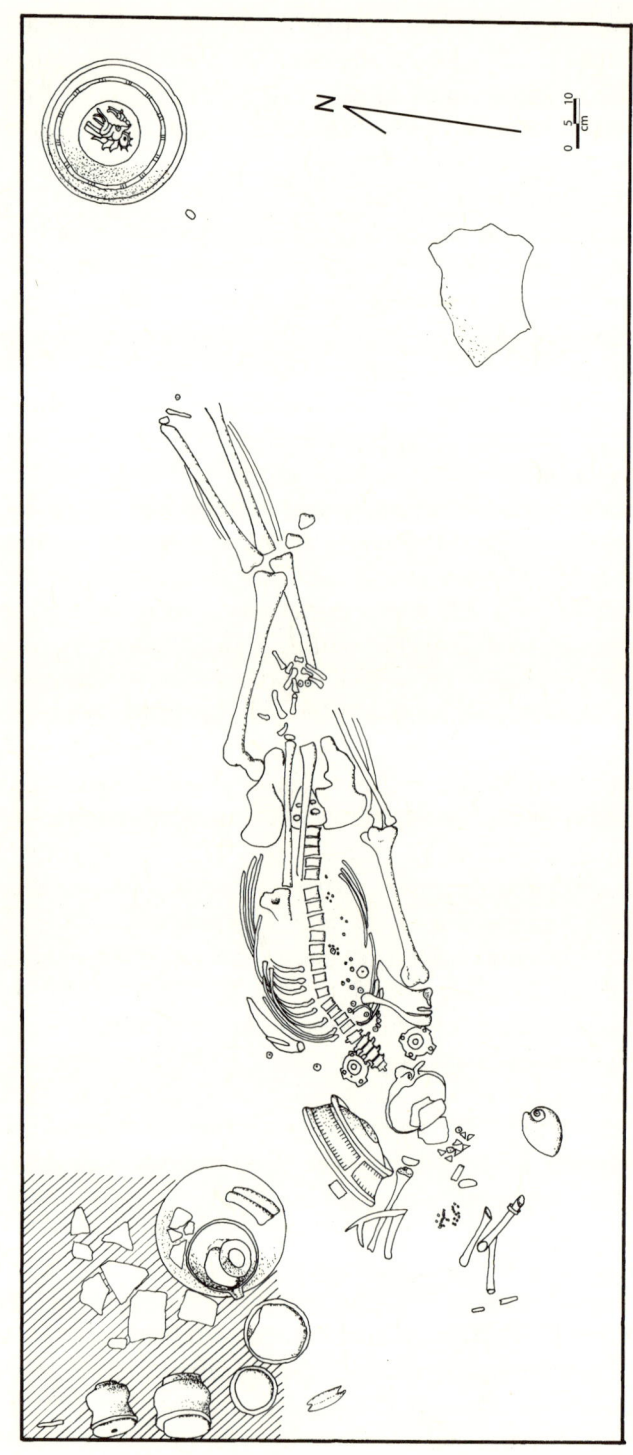

Fig. 7.1 Yaxuná's Burial 23. (Image created by Dave Johnstone.)

Tlaloc warfare (Freidel et al. 1990:13). Structure 6F-4-3rd probably represents a victory monument that housed the remains of the defeated lord, his family, and his royal regalia. At this point, we cannot say for certain whether this act of aggression represents an act of conquest followed by occupation or a local coup carried out with the support of a site in western Yucatán.

The changes in architecture and the accompanying break in the local dynastic sequence at Yaxuná during the Middle Classic period were as significant as the historically documented warfare and defeat suffered by Tikal at the hands of Caracol. At Yaxuná, though, the impact of these events on the ceramic assemblage was not as strongly felt. While locally produced wares ceased to be made and were replaced by types common to Oxkintok, these new types were not as well executed as those made at Oxkintok and did not cover the range of types produced at that site. Other ceramic types from eastern Yucatán were also present at Yaxuná, suggesting some trade with this region. This would suggest that while Oxkintok, or another site within the Oxkintok regional sphere, heavily influenced Yaxuná, this influence did not extend to direct control (Johnstone 2001).

During the Late Classic period, Yaxuná became incorporated into a larger political unit led by Cobá. Contrary to the initial belief that the Late Classic represented a hiatus at Yaxuná, this period was one of energetic building and reorganization. Coincident with this period was a change in ceramics and architectural styles, as well as termination deposits in residential and public buildings.

The Yaxuna III (Late Classic) ceramic complex represents a complete disjunction from those of the previous Yaxuna IIb complex, with earlier types being replaced by new ones. The most frequently occurring ceramic type at Yaxuná during the Late Classic period is Arena Red. Late Classic Yaxuná appears to have had very limited trade with its neighbors. The locally produced Arena Red is only found at Acanceh (Brainerd 1958) and at sites within the Greater Cobá polity (Robles Castellanos 1990), Xelhá (Canché 1992), and, to a limited extent, at Tancah (Ball 1982). Additionally, Cobá's most common type of this period, Lankin Impressed (and other related Batres group types), is not found in significant numbers at Yaxuná. Petén polychromes of many types reached Cobá in large numbers (Robles Castellanos 1990), but at Yaxuná, they are rare and limited to one type: Saxché Orange Polychrome. This suggests that the ceramic trade between Yaxuná and Cobá was not reciprocal. This system is an extractive economy, designed to enrich the capital at the expense of sites at the periphery.

Termination deposits accompany the change to the Late Classic pe-

riod. Structures 6E-53 and 6E-120, which Freidel and Suhler (1999) interpret as dance platforms or accession buildings, were carefully filled
in with Yaxuna III midden material. This careful, or venerating, termination is in marked contrast to the violent termination that was
carried out against the elite residences of Structures 5E-52 and 5E-75.
These residences, linked by Sacbe 8, were the largest and, in the case
of 5E-52, most elaborate residential structures excavated at Yaxuná.
Structure 5E-52's molded-stucco façade was destroyed and buried by a
layer of white marl (Freidel et al. 1990). The interior rooms contained a
thick layer of deliberately smashed ceramics of the Yaxuna III complex.
Structure 5E-75–1st had its walls stripped to their foundations and was
covered by a Late Classic residence reoriented to the north (Shaw and
Johnstone 2001).

The architectural styles present during the Late Classic period like-
wise show a disjunction with previous Early Classic architecture. The
ceremonial architecture associated with this phase at Structure 5F-3 is
quite similar to that of Late Classic Cobá, particularly to Cobá's Structures B-1 and C-1, which display battered terraces with inset rounded
corners (Thompson et al. 1932:32, 81). Additionally, thin-inclined
slabs with few spalls replace the corbelled vault composed of roughly
shaped thick rocks with many spalls.

An even clearer statement of takeover and rule is seen in the Late
Classic construction of Sacbe 1 connecting Yaxuná to Cobá. Sacbe 1
probably served multiple functions, including aiding the flow of commodities, permitting the swift transport of important messages, helping administrators travel between sites, and serving as a defensive
infrastructure by allowing armed forces to be relocated swiftly to crisis
areas. Cobá may have found the sacbe necessary as a means to forcibly
include this new area into its territory. Both basic forms of monumental architecture (temples and sacbeob) demonstrated the ability that
the political leaders had to control and extract labor from the populace
(Kurjack 1977:219). The 100 km sacbe not only provided a powerful
symbol of the relationship between the two sites, it also provided the
means to rapidly deliver the mandate of Cobá's rule to subject peoples
at Yaxuná.

Cobá's influence was further seen in the reorganization of settlement within the site. The construction of a large internal roadway,
Sacbe 5, helped reorient the site axis from north-south to east-west.
Additionally, the tallest pyramid at the site, Structure 5F-3, was reno-
vated and reoriented to face Sacbe 1's terminus. A new palace structure, 6F-8, was built on the southern edge of the North Acropolis, also
oriented towards the end of Sacbe 1.

The Terminal Classic period at Yaxuná begins around AD 750 with a less abrupt transition. New Puuc-style Florescent architecture is introduced, with some buildings constructed in a hybrid style of load-bearing and veneer masonry as seen at Dzibilchaltún. Many residences and several structures on the North Acropolis and the ballcourt were built or modified at this time, demonstrating that construction was also carried out on a large scale. Structure 6F-68, a council house on the North Acropolis, was built in a modified Puuc style with an elaborate carved facade (Ambrosino 2003; Suhler and Freidel 1993).

Most of Terminal Classic Yaxuná's ceramics can be included in the Western Cehpech ceramic sphere. The Yaxuna IVa ceramic complex represents a gradual evolution of ceramic types from the preceding Yaxuna III complex. During this period, although a large number of types and varieties are present, only a limited number of standardized forms are used (Johnstone 2001; R. Smith 1971). This may represent either a limited number of ceramic producers or a heavily elite-controlled system of production. Trade wares from the Petén cease to be imported, and new Thin Slate wares from western Yucatán take their place. The introduction of Puuc architecture and associated trade wares suggests that this region had a strong influence over Yaxuna at this time.

At the end of the Terminal Classic period (ca. AD 900), Yaxuná experienced a more catastrophic change in its political fortunes as the site was sacked. In spite of the apparent boom during the time the Yaxuna IVa complex was in use, there is evidence of an increasing threat during the Terminal Classic period. The best support for a state of hostility existing between Yaxuná and another site is the construction of an ad hoc fortification around the perimeter of the North Acropolis that contains a number of restricted entryways and overrides some Terminal Classic residences (Ambrosino and Manahan 1998). Within the wall are a number of residential foundation braces (Shaw and Johnstone 1996) similar to the "siege structures" at Dos Pilas (Demarest 1993). Ultimately, this fortification was not successful, and there is evidence of violent termination activities outside the wall at the ballcourt plaza (Johnstone 1994) and inside the wall at Yaxuná's council house, Structure 6F-68 (Ambrosino 2003), including the destruction of floors, vaults, and monuments, the burning of structures, the desecration of burials, and the deliberate breakage and scattering of ceramic vessels. The clue to the protagonist lies in this breakage and scattering.

The Yaxuna IVb ceramic complex abruptly replaced the Yaxuna IVa complex. This new complex consists of Sotuta ceramics that are most strongly associated with Yaxuná's nearest large neighbor, Chichén Itzá.

Prior to this disjunction, no Sotuta ceramics were found at Yaxuná. As the two sites, only 19 km apart, were contemporaneous in the Terminal Classic period, the absence of Sotuta ceramics from the Yaxuna IVa complex is surprising. The lack of trade during Yaxuna IVa suggests a powerful barrier prohibiting the free exchange of goods. Sotuta ceramics appear at Yaxuná for the first time directly associated with termination deposits marking the deliberate destruction of buildings (Johnstone 2001).

The nature of the relationship between Yaxuná and Chichén Itzá is suggested to have been one of hostility. The oral history of the *Chilam Balam* (Roys 1933) reports a series of battles between the two sites resulting in the payment of tribute to the Itzá. This suggests that Yaxuná, also known as Cetelac in the *Chilam Balam*, was the loser in this military campaign. The challenge is to corroborate the ethnohistoric accounts of warfare and to demonstrate the relationship between this military event and its archaeological correlates, including the change in ceramic complexes.

Following the Itzá conquest and destruction of Yaxuná, few new buildings were constructed. Only one, Structure 6F-9, contained Sotuta ceramics and cache material in its construction fill. The absence of significant reconstruction following this war-related event is in marked contrast to previous periods of political upheaval. After this episode, Yaxuná does not seem to have regained its independence and suffered a substantial decline in population. Yaxuná's proximity to Chichén Itzá may have contributed to its more complete incorporation within that polity, with its population carried off to Chichén Itzá to give tribute in labor (Shaw 1998).

Discussion and Conclusions

Traditionally, the northern lowlands have been viewed as having followed a different political developmental trajectory than that seen in the southern lowlands. Recent research has begun to illustrate that some of those processes that affected the south were also present in the north. Almost from the beginning, the north has evidence for divine rulership in the form of royal kings. These kings did not rule in splendid isolation but traded widely and were aware of, and participated in, political relations with cities in the south. The presence of glyphic texts at Yo'okop mentioning Calakmul, and the introduction of Teotihuacanoid elements to art and architecture of sites such as Yaxuná and Oxkintok (Rivera Dorado 1991) by way of Tikal attest to the influence that these two "superstates" had in the northern lowlands.

Research has rejected the Yaxuná Project's first hypothesis of Cobá having constructed Sacbe 1 as a response to Chichén Itzá aggression. Sacbe 1 was built in the Late Classic period, before Chichén Itzá was a significant power. Instead, the sacbe was used as a means of expanding the borders of the Cobá state. If the other road networks of the north were also constructed during the Late Classic, then this period would be one characterized by active expansion of certain polities, with newly incorporated areas integrated by means of a road system over which armies could rapidly deploy.

Ultimately, this system appears to have broken down in the Terminal Classic period, with many sites—including Uxmal, Aké, Muna, Cuca, Chacchob, Yaxuná, Chichén Itzá, Ek Balam, and Dzonot Aké (Ambrosino and Manahan 1998; Barrera Rubio 1985; Bey et al. 1997; Kurjack and Andrews 1976; Ruz L'huillier 1951; Webster 1979)—being fortified during this period. At Yaxuná, the second hypothesis of ethnohistorically reported conflict between Chichén Itzá and Yaxuná has been supported by evidence for Chichén Itzá–associated pottery appearing for this first time in direct association with termination deposits associated with the sacking of the city. Whether this policy resulted in the fortification of other sites at this time remains the subject of future research.

The use of several independent lines of research, known as the conjunctive approach, as well as the inclusion of more quantitative data and statistical analyses, are allowing modern archaeologists to better test hypotheses about Classic Maya politics. In spite of the insights provided by these methods, many questions remain for further research in the north. Why was the north able to continue, even peak, when many southern sites collapsed? Why didn't many northern sites leave the same political signatures (epigraphic texts) as southern sites did? What role did major southern sites play in northern politics? Did larger extra-regional alliances exist? What impacts did political upheavals have on non-elite Maya? To what degree were political leaders involved in the economy? In order to address these questions thoroughly, researchers need to focus on a better integration of site center and periphery studies, rather than solely focusing on monumental architecture in site cores. Additionally, there is a continued need for a better placement of sites in their regional context, not just the creation of site-specific sequences, for an improved understanding of artifact (especially ceramic) manufacturing and distribution patterns and an ever-present need for excavation of more in-context materials to allow dates to be assigned to specific occupation episodes.

Editor's Note

Ceramic type names are taken from modern Maya pueblos and so reflect current spelling and therefore may include accents (such as Saxché Orange Polychrome). However, ceramic sphere names are assigned under an older orthography that does not include accents. Therefore, while the site of Yaxuná should be written with an accent, the ceramic complexes from the site (such as Yaxuna IIb) should not have accents. These conventions have been used in the preceding text.

8

Ichmul de Morley and Northern Maya Political Dynamics

J. Gregory Smith, William M. Ringle, and Tara M. Bond-Freeman

The northern Maya lowlands offer the archaeologist a remarkable laboratory within which to address issues of political organization. The issue of how best to characterize the political organization of the Maya is an old one and is still the subject of much debate (Fox et al. 1996). Several Mayanists propose that Maya polities were strong, centralized unitary states (A. Chase and Chase 1996; D. Chase and Chase 1992; Culbert 1991; Folan 1992; Marcus 1993, 1995), while another group argues that the Maya are best characterized as smaller-scale, decentralized, segmentary states (Adams and Smith 1981; Ball and Taschek 1991; Coe 1961; Demarest 1992; Dunham 1990; Fox 1987; Houston 1992).

As noted by Bey (this vol.), most discussions of ancient Maya political organization reflect a perspective biased toward the southern lowlands. Practically all of the works cited above in the ongoing unitary vs. segmentary debate are based on work carried out at southern sites. There are some notable exceptions to this pattern. While Garza Tarazona de González and Kurjack (1980) did not use the segmentary state model per se, their observation that secondary centers often rivaled the size of their capitals led them to conclude that there was a lack of centralized power in northern lowland polities. More recently, the segmentary state model has been used specifically to interpret the Ek Balam polity (Bey and Ringle 1989; Ringle and Bey 1992, 2001) and the polities of the Puuc region (Dunning 1992; Dunning and Kowalski 1994). Ball (1994) has suggested that, while most northern lowland polities were segmentary states, Chichén Itzá was a centralized unitary state, although others disagree (Ringle and Bey 1992; Schele and Freidel 1990). Despite these contributions, the northern Yucatán in general is still underrepresented in the discourse concerning Maya political organization.

Another bias in the study of Maya political organization has been an over-emphasis on a single site as the unit of analysis. Mayanists have generally underutilized a regional approach, yet it is understood that many interesting components of political systems can only be studied by employing a regional perspective (Fish and Kowalewski 1990). The few regional surveys that have been carried out are usually limited to those areas where lower vegetation permits full-coverage surveys (de Montmollin 1989, 1995; Kepecs 1999; Webster 1985). Due to the thick vegetation over most of the lowlands, several projects have conducted surveys between large sites by means of transects of varying widths. While transect projects are less comprehensive than 100 percent polity-wide surveys, they at least approach the study of the Maya from a regional perspective (Dunham 1990; Ford 1990; Puleston 1983; D. Rice and Rice 1990; Vlcek and Fash 1986). With only a handful of these kinds of projects for the entire Maya area, it is clear that there is a lack of regionally oriented approaches aimed at understanding ancient Maya polities.

The Ek Balam Project (Ringle et al. 2003, 2004) was conceived to partially remedy this need in the north. At the time our fieldwork was conducted, the only other projects in the northern Maya lowlands with an explicitly regional focus were Dunning's (1992) study of the eastern Puuc area and Fedick and Mathews's work in the Yalahau region (Fedick and Taube 1995; J. Mathews 1998). Garza Tarazona de González and Kurjack's (1980) archaeological atlas of Yucatán (*Atlas arqueológico del estado de Yucatán*) provides an invaluable macro-regional perspective within which to situate these more focused studies. Our research design emphasized settlement study at three levels of analysis: within the urban zone of Ek Balam, within the immediate sustaining hinterlands (Houck, this vol.), and finally, between Ek Balam and its neighboring peer polities, the subject of this chapter. Given the magnitude of a 100 percent study of intersite settlement, we selected a 20 km wide transect between Chichén Itzá and Ek Balam for intensive sampling (fig. 8.1). J. Gregory Smith (2000) carried out the greater part of the fieldwork within this transect. In this chapter, we would like to consider one site selected for more intensive work, Ichmul de Morley. This site is not only one of the largest within the transect but also lies nearly midway between its two larger neighbors, thus providing an interesting test case for models of political organization and boundary construction. While several Mesoamericanists have studied boundaries (Beekman 1996; Dunham 1990; Gorenstein 1985; Kowalewski et al. 1983; Marcus 1984; Redmond 1983; Silverstein 2001), our work is the most detailed boundary analysis available for the northern Maya lowlands.

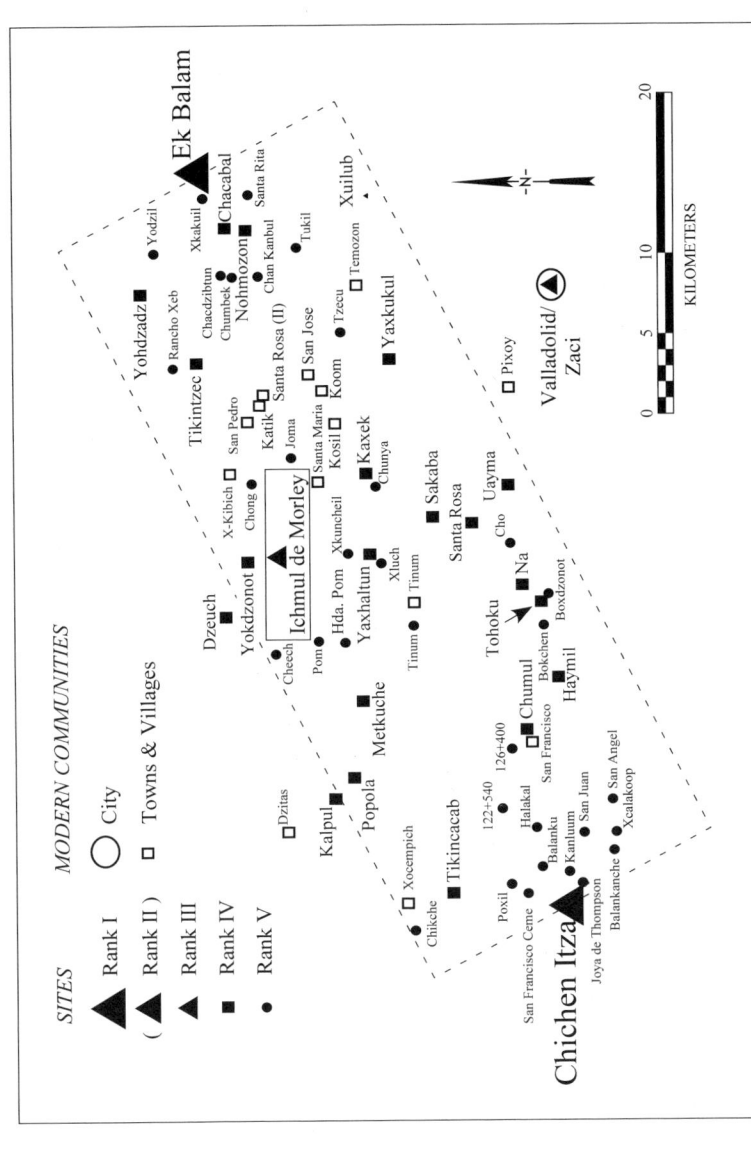

Fig. 8.1 The Chichén Itzá–Ek Balam transect showing the location of Ichmul de Morley. (Map created by J. Gregory Smith.)

Political Organization and Boundaries

Two models of political control help conceptualize the strategies that political leaders in polity capitals use to deal with their boundaries. In some cases, polities establish a defined territory and attempt to maintain this territory by demarcating its boundaries. Political leaders at the capital exercise what Hassig (1992) has called territorial control and tend to distribute their power evenly over the extent of the polity. Local elites in this case are forcibly removed and replaced by administrators from the conquering polity, the new administrators forming part of an organized bureaucracy that acts to consolidate these new territories. Boundaries in polities using this strategy of political control are often demarcated by means of garrisons or fortifications. From the perspective of the ruling elite at the capital, the territorial strategy of control is extremely costly.

Hicks (1991) offers a somewhat similar distinction between the *gift* and *tribute* economies of the Aztecs, the latter being characteristic of forcible extraction and the former being characteristic of a network of unequal prestations. He too makes the point that tribute extraction must be backed by the threat of force and hence is more costly and risky. This mode of political control requires a relatively high-energy investment to conquer a territory, install administrators to consolidate it, establish garrisons and military outposts along its boundaries, and then to maintain this control apparatus indefinitely. Given this, why would political leaders choose to employ the territorial strategy of control? The high costs in establishing a territorially based polity are matched by high benefits. Since control is direct, the polity is able to extract more surpluses from the conquered populace: less goods and services are siphoned off by local elite and instead go directly to leaders at the capital.

In what Hassig (1992) has called the hegemonic strategy of political control, political leaders at the capital control peripheral settlements indirectly. Instead of installing officials from the capital to directly administer outlying settlements, local elites are left in place and, aside from tribute demands, are unmolested. Political power tends to be a function of distance, resulting in ambiguous boundaries between capitals. The segmentary state models developed by political anthropologists (Geertz 1980; Southall 1956, 1988; Tambiah 1977) suggest that leaders in these kinds of polities are less concerned with maintaining a defined territory than in establishing alliances of varying strengths. Buffer zones often form between polity capitals, and if they are controlled at all, communities within such zones are administered indirectly. Dunham (1990) has argued that elites in sites halfway between capitals of hegemonic states are especially likely to remain autonomous since they are maximally distant from the power of either capital. With little political infrastructure and lacking stable boundaries, hegemonic states tend to allow peripheral settlements to more easily break away from the capital and have difficulty keeping local political leaders permanently subordinate. The success of the polity instead often depends on the personal charisma of the paramount ruler.

Given these inherent problems, the question can again be asked: why would political leaders in some cases opt for the hegemonic strategy of political control? The answer is that it is a low-cost, low-risk strategy: polity capitals do not invest much energy in conquering or maintaining control over the peripheral populace. Although not controlled directly, local elite are subservient to the capital and must pay tribute. If the paramount leader at the capital can establish hegemonic relationships with many neighboring polities, a substantial amount of tribute can flow into the capital. Typically, however, revenues are substantially diminished by a process akin to tax farming, in which portions of tribute collections are extracted at each level of the hierarchy.

To place these two models of territorial control in the context of the debate concerning Maya political organization: unitary states tend to employ territorial control, while segmentary states are often associated with hegemonic control. That being said, we should be wary of assuming that dimensions of political systems necessarily align themselves according to the unitary-segmentary dichotomy. Furthermore, both the unitary-segmentary and hegemonic-territorial pairings should not be considered as mutually exclusive possibilities, but rather as the poles of a continuum. For instance, polities do not deal with all of their boundaries in a uniform way, often employing both strategies of control. The Aztecs, for instance, generally controlled inner provinces hegemonically and outer provinces territorially.

Ichmul de Morley

As noted above, the location of Ichmul de Morley makes it a particularly interesting site from which to examine these issues (see fig. 8.1). Ceramic analysis demonstrates that Ichmul experienced its florescence during the Late/Terminal Classic period, at the same time as its two larger neighbors, with only minor occupations before and after (Bond-Freeman et al. 1999). The site first appeared in the archaeological literature (and received its name) following a 1918 visit by Sylvanus Morley (1919:figure 1), during his search for hieroglyphic inscriptions in Yucatán. The site remained uninvestigated by archaeologists until its registry in the Yucatán Atlas Project (Garza Tarazona de González and Kurjack 1980). Members of the Cupul survey (A. Andrews, Gallareta Negrón, et al. 1989) briefly visited Ichmul in 1988 and noted small quantities of Sotuta ceramics. In 1990, Gallareta Negrón, Ringle, Hanson, and Hartzell-Scott (Gallareta Negrón 1991) mapped the civic architecture and produced the first site map (fig. 8.2), as well as recovering another small surface collection of pottery. The data presented in this chapter come from a six-week field season during the summer of 1997. This project concentrated on a 30 ha settlement survey and the excavation of several elite and commoner structures (Bond-Freeman et al. 1999; Ringle and Smith 1998; J. Smith et al. 1998).

Questions/Problems

How might the strategies of political control exercised by Chichén Itzá and Ek Balam be detectable at Ichmul de Morley? Redmond (1983) has suggested that the presence of imperial symbols, garrisons, and/or cessation of contacts between a subjugated area and polities other than the imperial capital would reflect the consolidation of political power by a capital. If Chichén Itzá or Ek Balam employed the territorial strategy of political control, then a reasonable expectation would be that Ichmul should affiliate closely with its capital and perhaps contain fortifications or other evidence of military garrisoning. On the other hand, if Chichén Itzá and Ek Balam both used the hegemonic strategy, Ichmul would be left relatively autonomous. An autonomous Ichmul might incorporate features of both polity capitals as well as having some attributes of its own.

This easy generality is less than satisfactory for several reasons, however. Political contacts outside of a centralized system might diminish, but economic contacts might continue unabated. Thus, it is important to determine how foreign material remains end up at putative border sites. Furthermore, it is often difficult to know how symbols are being utilized. Given the fusion of political and ideological representation typical of archaic states, symbols may be expressions of either imperial control or of *hegemonic* participation in looser ideological networks. The *theater state* variant of the segmentary model (Geertz 1980), in fact, emphasizes that display of ideological symbols is crucial to political survival. Thus, imagery cannot be assumed to be transparent.

Several classes of data are considered in evaluating the models of political control outlined above. For each class of data—civic layout,

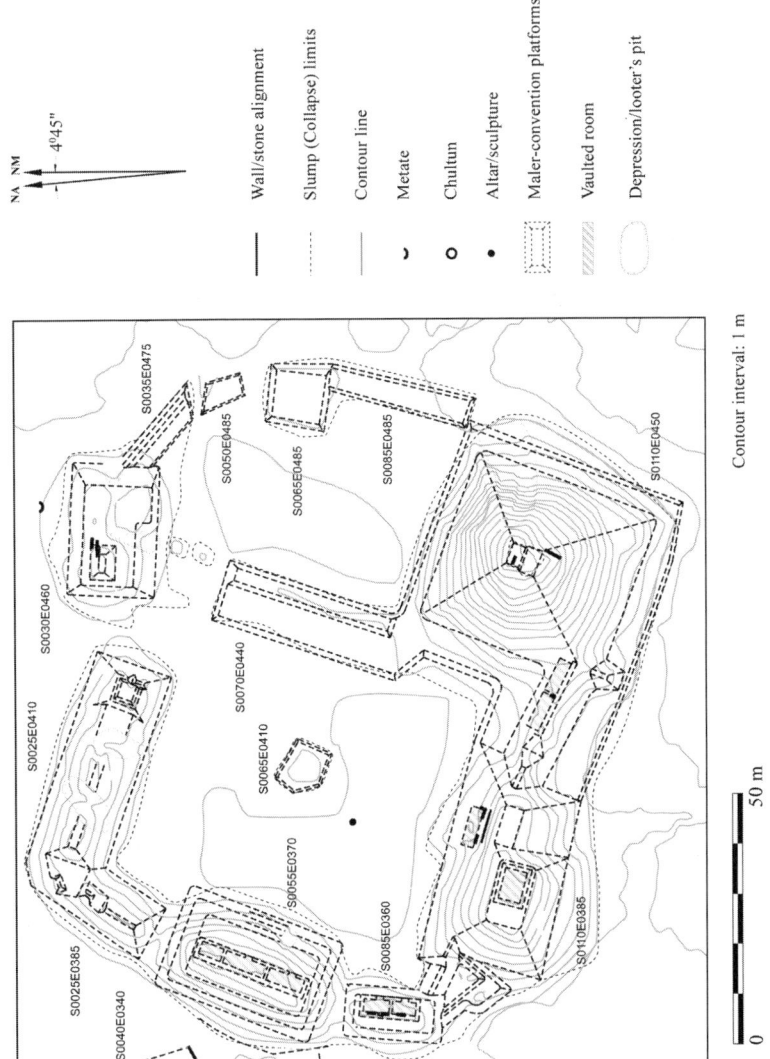

NA NM

4°45"

Wall/stone alignment

Slump (Collapse) limits

Contour line

Metate

Chultun

Altar/sculpture

Maler-convention platforms

Vaulted room

Depression/looter's pit

S0035E0475

S0050E0485

S0065E0485

S0085E0485

S0110E0450

S0030E0460

S0070E0440

S0025E0410

S0065E0410

S0055E0370

S0025E0385

S0085E0360

S0040E0340

S0110E0385

Contour interval: 1 m

0

50 m

Fig. 8.2 The civic center of Ichmul de Morley. (Map created by William Ringle and J. Gregory Smith.)

architectural styles, iconography, ceramics, and obsidian—the differences between Chichén Itzá and Ek Balam are first outlined, followed by the evidence from Ichmul de Morley. Distinguishing the presence of Chichén Itzá and Ek Balam at Ichmul de Morley is fortunately relatively straightforward because the two larger centers differ systematically in each of the data classes.

Civic Layout. The layout of Chichén Itzá features a large terrace called the Gran Nivelación supporting a core group of public structures, the so-called "New Chichén." This central plaza is relatively open. Although the Great Ballcourt bounds the west side and the Temple of the Warriors–Mercado complex dominates the east, the north and south edges of the Gran Nivelación are devoid of major architecture. Several other large architectural groups such as *Old Chichén* and the Far East Group are tethered to the site center by raised causeways or *sacbeob*. Perhaps the most distinct feature of Chichén's civic layout is the placement of its major civic structure, the El Castillo radial pyramid, roughly at the center of the Gran Nivelación.

The civic center of Ek Balam is focused on a main plaza bounded on three sides by monumental buildings and to the south by several lesser, but still impressive, structures. Ek Balam's main civic structure, GT-1, delimits the north edge of the main plaza. GT-1 is a huge multi-level acropolis supporting several *plazuelas* and superstructures and probably was at least in part a war temple. Instituto Nacional de Antropología e Historia (INAH) excavations in this structure following the close of our project have provided a wealth of new information (Vargas de la Peña and Castillo Borges 1999, 2001), including many hieroglyphic texts with dates (Grube et al. 2003; Vargas de la Peña et al. 1999). This enormous building is flanked on the west by a range structure (GT-2) fronting a wide raised terrace, presumably for assemblies and festivities, and on the east by another large multistoried palace-like structure (GT-3). The civic layout of Ek Balam is quite different from Chichén's, both by virtue of differences in access to public spaces and by the types of structures found around the main plazas.

What would seem to be the main plaza at Ichmul de Morley is bounded on three sides by vaulted structures limiting access to its central plaza (see fig. 8.2). In this regard, the civic center of Ichmul de Morley is more similar to the main plaza of Ek Balam, but the arrangement of structures enclosing the plazas differs significantly. Most of the vaulted structures are simple structures placed upon low platforms, perhaps elite residences. Although poorly preserved, the 17 m high main mound at Ichmul is clearly not a radial pyramid in the tradition of the Castillo, but it does not much resemble the GT-1 acropolis of Ek Balam either. Furthermore, this mound is located off-center near the southeast corner of the plaza. Its staircase, presumably descending the northern face, would not even have been accessible from the main plaza. Instead, a long, low platform with a bench along its eastern edge extends northward from the pyramid to define the eastern side of the plaza.

In this respect, these two Ichmul structures resemble some of the lesser civic complexes of Ek Balam, such as the Grupo Suroeste, and those at transect sites such as Chumul. At Ichmul, it seems as if the main mound and the attached long buildings actually formed a small plaza to the east of the main plaza. In sum, the civic layout of Ichmul has little resemblance to Chichén Itzá and only slightly more to Ek Balam. Rather than being an integrated complex uniting temples and elite residences, it seems that at Ichmul the two were contiguous but spatially distinct.

Architectural Styles. The distinct architectural style present at Chichén Itzá is commonly referred to as Modified Florescent. One common feature of this style of architecture is the use of columns to create large interior spaces, such as is found in gallery-patio structures and colonnades. Other Modified Florescent traits include radial pyramids, serpent columns, I-shaped ballcourts, and a variety of non-Maya architectural ornaments. Chichén Itzá also has a style of domestic architecture that can be called "File Houses." These dwellings often have a front and back room (arranged like the files in a file cabinet) with a single exterior doorway, often with columns.

Ek Balam has no Modified Florescent architecture. Instead, the form and ornamentation of its buildings reflect several styles found at Late/Terminal Classic sites across the northern lowlands. Recent consolidation efforts of GT-1 by the INAH have revealed a platform fronted by numerous vaulted rooms and the extensive use of modeled stucco as decorative elements (Vargas de la Peña and Castillo Borges 1999). Both there and in our own excavations, traces of Chenes-style and Puuc-style construction may be found. The two ballcourts at Ek Balam do not have end zones, as at Chichén Itzá, and are instead open-ended like the majority of other courts in the north. The domestic architecture at Ek Balam is dominated by the two- or three-room perishable structures, each room of which typically opens to the exterior. Vaulted residences outside the civic complexes are rare.

The civic architecture at Ichmul lacks any candidates that could be considered Modified Florescent. Our fieldwork at Ichmul has failed to identify Chichén Itzá–inspired buildings such as gallery-patio complexes or I-shaped ballcourts. Architectural elements commonly found at Chichén Itzá, such as colonnades, are also absent. One cylindrically shaped stone was found in the main plaza, which may be a column drum. However, since columns consist of many stacked drums and no other such stones were found, a more likely explanation is that it functioned as an altar of some sort. In terms of domestic architecture, the majority of the houses we mapped at Ichmul resemble those of Ek Balam. A minority were of a type we have labeled "open-front" houses, perishable dwellings featuring two rooms connected by a rear frame brace wall, presumably leaving an open central room between them. Examples have also been found at other sites in the Chichén Itzá–Ek Balam transect and at Kumal, but their greater frequency at Ichmul may be significant.

Ceramics. The ceramics associated with Chichén Itzá are collectively known as the Sotuta ceramic complex (R. Smith 1971). Robert Smith originally defined the Sotuta complex as being subsequent to the Cehpech complex; the ceramics are found across virtually all of northern Yucatán save for Chichén Itzá and its dependencies. As is well known, the chronological relationship between the Sotuta and Cehpech spheres has been debated for decades with most scholars now arguing for at least partial overlap between the two (see summary and references in Ringle et al. 1998). Perhaps the most notable difference between the Sotuta and Cehpech complexes is the presence of tradewares such as Tohil Plumbate and Silho Fine Orange in the former and the absence of these fancy vessels in the latter. Although the local wares of the two complexes are technically similar and often hard to distinguish, recent excavations at Chichén Itzá and near outliers have recovered nearly pure Sotuta deposits (P. Anderson 1998a, 1998b; Pérez de Heredia Puente 1998).

The Ek Balam Project has analyzed over a quarter of a million sherds at the site and its environs and has conclusively demonstrated that the site participated in the Cehpech ceramic sphere during the Late/Terminal Classic periods (Bey et al. 1998). Only 145 Sotuta sherds have been recovered at Ek Balam, less than .01 percent of the Late/Terminal Classic assemblage. The Cehpech assemblage at Ek Balam resembles the ceramics found across the northwest plains and the Puuc area more than those of Cobá, although like most such assemblages it has its own idiosyncrasies.

The ceramics at Ichmul de Morley have an interesting mix of types from both the Cehpech and Sotuta complexes (Bond-Freeman et al.

1999; Ringle et al. 2003; J. Smith 2000). Looking at the Late/Terminal Classic ceramic assemblage at Ichmul, of the 13,438 identified sherds recovered, 12,005 (89.3 percent) were Cehpech and 1,412 (10.6 percent) were Sotuta. The Cehpech fraction is probably inflated, since it includes all sherds identified as "eroded slateware," some percentage of which are probably Sotuta. However, 3,938 sherds could be positively placed in the Muna Slate group, but only 625 sherds belonged to the corresponding (Sotuta) Dzitas Slate group. With regard to unslipped sherds, 2,272 belonged to the Chum group versus 553 of the Sisal group. There is thus a fairly consistent 4–6-fold predominance of Cehpech versus Sotuta types. It is important to note that there is no distinct Sotuta horizon at Ichmul: Sotuta ceramics were found associated with Cehpech on the surface, in construction fill, and in midden contexts throughout the site. Thus, the two complexes, as manifested at Ichmul de Morley, are mostly, if not entirely, contemporaneous. Moreover, there were many examples of pots at Ichmul with combined attributes of both the Sotuta and Cehpech spheres. For example, several *molcajetes* (grater bowls)—a Sotuta form—had the paste and slip of Muna Slate, a very common Cehpech type.

Since there were several examples of these kinds of hybridized vessels at Ichmul, it is clear that this site was not wholly dependent on either capital for its ceramics and had its own ceramic tradition in some ways distinct from both Chichén and Ek Balam. The fact that Sotuta sherds are scattered widely, but by no means ubiquitously, across the site in both elite and commoner domestic contexts indicates that these ceramics were not the result of special events such as termination rituals, as has been argued at Yaxuná (Freidel 1992). Our elite structures were not as well sampled as might be desired, but the levels of Sotuta sherds there were even lower than outside the main plaza. Not one of our units had more than a few sherds from the same vessel, and in all cases, the Sotuta fraction was minimal.

Obsidian. Recent excavations conducted by members of an INAH project at Chichén Itzá have recovered some 2,745 pieces of obsidian that were subsequently analyzed by Braswell (1997, 1998). Braswell's study found that 75 percent of this assemblage came from sources in Mexico, with the two most utilized sources being Ucareo (32 percent) and Pachuca (21 percent). The remaining 25 percent of the Chichén obsidian came from the Guatemalan sources of Ixtepeque, El Chayal, and San Martín Jilotepeque.

Obsidian collected at Ek Balam by INAH ($n = 198$) and also analyzed by Braswell show a very different pattern than at Chichén Itzá. A full

Table 8.1 Pachuca Percentages for Chichén Itzá, Ek Balam, and Ichmul de Morley

Site	n	Pachuca	Percent
Chichén Itzá	2,745	577	21
Ek Balam	198	3	2
Ichmul de Morley	38	8	21

98 percent of the Ek Balam obsidian collection comes from Guatemala, with almost all of it sourced to El Chayal. Only 2 percent ($n = 3$) of the Ek Balam obsidian came from Pachuca, Hidalgo. The obsidian collected by the Ek Balam Project of Ringle and Bey matches this pattern. The obsidian data from Ek Balam make it clear that this polity simply did not have or did not want access to Mexican obsidian.

During the 1997 season, thirty-eight pieces of obsidian were recovered at Ichmul de Morley. A complete analysis of the Ichmul material is pending, and therefore we concentrate discussion here on the most easily identifiable source—the green obsidian of Pachuca, Hidalgo. Eight pieces of obsidian in the Ichmul assemblage were visually sourced to Pachuca. Table 8.1 presents the comparative frequencies of Pachuca obsidian from each of the three sites. Although the Ichmul sample size is admittedly small, its Pachuca percentage is identical to that of Chichén Itzá and very different from that of Ek Balam.

Iconography. The iconography of Chichén Itzá is quite distinct from every other Maya polity, including Ek Balam. The term "Toltec" is often used to describe Chichén's iconography and is used generically to differentiate it from Classic Maya iconography (with which it is often combined). As is the case with most other scholars, we use the term Toltec as purely a stylistic label and not as a cultural-historical one (for a current interpretation of the relationship between Chichén Itzá and the Toltecs at Tula, see Cobos, this vol.). Common attributes of the Chichén Toltec iconographic program include bas-relief sculptures and murals with complex narrative scenes featuring numerous figures, serpent columns, atlantean figures, representations of jaguars and eagles, non-Maya glyphs and costume elements, and *chacmool* sculptures (Taube 1994; Tozzer 1957).

Turning to Ek Balam, the chief parallels with Chichén are traits found at other Maya sites as well, such as God K capstones and Chac masks. The stelae there depict single regal figures dressed in typical Late

Classic Maya pose and costume (Ringle et al. 2003). There are no overtly Toltec images at Ek Balam, but some of the costume elements found on free-standing sculpture we found at the site do find parallels with Chichén, as do the large-scale reliefs recently found on the low terrace of GT-1 by INAH. It should be pointed out that many, if not all, of the Maya iconographic traits at Ek Balam are also found at Chichén Itzá. In effect, it appears to us that the iconography of Chichén Itzá shows an overlay of foreign traits on a Maya substrate, and Ek Balam exhibits only this Maya foundation.

The iconographic evidence from Ichmul de Morley consists of two sculptured panels first reported by Morley (1919). Photographs of the panels were not published until much later (Proskouriakoff 1950), and more recently, Greene Robertson (1993) has made available rubbings of both pieces. Panel 1 (fig. 8.3a), now in the Museum of Anthropology in Mérida, depicts two opposing ballplayers leaning down in the Classic ballplayer pose to strike a ball between them with their hips. Both wear elaborate headdresses, are bare-chested, and wear hip protectors. Glyph blocks run along the outer perimeter of the panel, and two L-shaped blocks are found between the two figures, above the ball. The broken Panel 2 (fig. 8.3b), now in the lobby of the Mérida Misión Hotel, also depicts ballplayers. It is quite similar to Panel 1 in that the two ballplayers with elaborate headdresses oppose each other in the hip-shot pose. Above them are two inverted L-shaped glyph blocks.

These two panels are quite similar to southern lowland ballplayer scenes (e.g., panels from Site Q/La Corona and Yaxchilán) and wholly distinct from the Toltec ballplayer scenes exemplified by the panels of the Great Ballcourt at Chichén. In addition to the pose of the players, the headdresses of the players incorporate bird and deer heads, as do many examples from the southern lowlands. Furthermore, the ball of Panel B incorporates a head and numerical coefficient, another common southern trait not characteristic of the Great Ballcourt sculpture.

Strangely, we found no ballcourt at Ichmul, suggesting the games occurred elsewhere. Recent epigraphic work strongly indicates they took place at Ek Balam. Graña-Behrens (2002:250–52) and others (Grube et al. 2003:II-30) have tentatively identified the name of Ukit Jol Ahkul, a king of Ek Balam, in association with the right player on Ichmul Panel 2, dated to 10.0.0.0.0 (AD 830). Ichmul Panel 1 remains undated, but these authors concur in identifying the Ek Balam king Ukit Kan Le'k along the left edge of the stone, unfortunately not clearly associated with either figure.

a

b

Fig. 8.3 Iconography at Ichmul de Morley: a) Panel 1. (Greene Robertson 1993: Rubbing 20024); b) Panel 2. (Greene Robertson 1993: Rubbing 20025). (Rubbings by Merle Greene Robertson ©Pre-Columbian Art Research Institute, 1995; used with permission.)

Discussion and Conclusions

When the evidence from Ichmul de Morley is examined (see table 8.2), it is clear that this site is not easily assigned to either Chichén Itzá or Ek Balam, suggesting neither site controlled Ichmul directly. The lack of an overt Chichén Itzá presence at Ichmul is especially noteworthy considering that Ichmul would have been a critical node on the route between Chichén and its port of Isla Cerritos (A. Andrews, Gallareta Negrón, et al. 1989; Kepecs et al. 1994). As Ichmul is the closest such node to Chichén, if the Itzá engaged an active policy of conquest warfare to secure this vital trade route, then one would expect this site to exhibit the most evidence for complete dominance and absorption in the Chichén polity. Not only is there no evidence of incorporation and infrastructural investment of the Itzá at Ichmul, there are no defensive walls and no evidence of an invasion. The lack of walls cannot be explained by Ichmul's small size, since some small sites in Yucatán do have walls (Webster 1978). The site of Yaxuná, even closer to Chichén Itzá than Ichmul, hastily erected defensive walls in response to Itzá aggression (Manahan et al. 1997).

Lest we be accused of portraying the rulers of Chichén Itzá and Ek Balam as being unable or unwilling to conquer nearby settlements, we do recognize cases of domination between larger capitals and secondary centers. The results of archaeological research at Yulá, a site located 5 km south of the center of Chichén Itzá, provides an interesting contrast with the findings at Ichmul de Morley. Work conducted by P. Anderson (1998a) revealed that Yulá is an example of a site that was very closely tied to Chichén Itzá. Nearly 97 percent (14,408 out of 14,930) of the Late/Terminal Classic ceramic assemblage is Sotuta with only a trace of Cehpech ceramics. Two lintels at Yulá feature iconography and hieroglyphic inscriptions that are very similar to lintels at Chichén Itzá, especially the Temple of the Four Lintels. P. Anderson (1998a:157) has suggested that Yulá functioned as a Chichén outpost and was occupied by members of one of its ruling lineages. Another example may be X'telhu, 29 km from Chichén, whose carved panels could represent conquests by Itzá warriors, though the specifics of these events are unclear (Robertson 1986). In a similar vein, the secondary center of Xuilub (Houck, this vol.) appears to have been thoroughly absorbed into the political orbit of Ek Balam.

Such results are not unexpected given the difficulties other archaeological projects have had in attempting to determine frontiers and borders on material grounds, but the evidence from Ichmul suggests a more complex situation. It seems to have been neither a simple out-

Table 8.2 Comparison of Datasets between Chichén Itzá, Ek Balam, and Ichmul de Morley

Class of Data	Chichén Itzá	Ek Balam	Ichmul de Morley
Civic layout	open plaza, main civic structure within plaza	closed plaza, main civic structure on north edge of plaza	closed plaza, main civic structure on southeast corner of plaza
Civic architectural styles	Modified Florescent	no Modified Florescent	no Modified Florescent
Domestic architecture	"file" houses	"row" houses, a few "open-front" houses	"row" and "open-front" houses
Ceramics	~100% Sotuta	~100% Cehpech	89% Cehpech 11% Sotuta
Obsidian	21% Pachuca	2% Pachuca	21% Pachuca
Iconography	"Toltec"	Classic Maya	Classic Maya

post of Chichén or Ek Balam, nor a part of what Kepecs (1997) calls "Itzá infrastructure." Instead, its differential participation in pottery and lithic distribution networks indicates that Ichmul de Morley may have been relatively autonomous, although probably with some sort of affiliation with its larger neighbors. This is further supported by the recent assertion that Ichmul de Morley may have had its own emblem glyph (Graña-Behrens 2002:252; Grube et al. 2003:II-21). Best seen on Ichmul de Morley Panel 2, Graña-Behrens argues that it has the same form as the Tikal emblem glyph (T569). Both panels from Ichmul depict two facing ballplayers and in both cases the opponent is a king of Ek Balam. To understand what these panels signify, it is important to understand how the ballgame functioned in mediating regional political relationships. Recently, Ringle (2004) has argued that ballgame rituals were central to the investiture rituals of lesser lords at Chichén Itzá. Such lords probably came from far beyond the Chichén polity and may have owed only ritual fealty to Chichén Itzá. Some suggestion that Ek Balam played a similar role as king-maker comes from the text of Ek Balam Stela 2. Just above its emblem glyph on the left edge is a compound that seems to read T32.86?:700. The latter glyph is clearly used in seating rituals at Palenque and may represent a zoomorphic throne in profile. A possible translation might be "holy seating/throne place." In other words, the lords of Ichmul probably received legitimacy from the overlord of Ek Balam but afterwards may have functioned relatively independently.

Our view of a fluid and loosely controlled boundary between Chichén Itzá and Ek Balam is consonant with ethnohistoric descriptions of the proto-historic polities of northern Yucatán. Several of the *Relaciones de Yucatán* (de la Garza 1983) indicate that tribute obligations owed by peripheral centers were nominal (i.e., archaeologically invisible) and that alliances seemed mostly to have involved military concerns. Quezada (1993) has also argued forcefully that these *batabilob* or *cuchcabalob*[1] were only occasionally territorially contiguous; in eastern Yucatán, the vectors of power radiating from Saci, Chichén Itzá, and Ek Balam crisscrossed each other in a complex fashion. Data gathered by Ek Balam Project (Ringle et al. 2003, 2004; J. Smith 2000) show that external influence at other regional sites reflects this complexity.

Although the mixed pattern at Ichmul may be due to our limited evidence, it may also reflect deeper problems in conceptualizing how material culture might pattern in border situations. In an overview on the topic, Lightfoot and Martinez (1995) remind us that our preconceptions of how borders and frontiers might be manifested in the archaeological record are often colored by colonialist assumptions concerning the transmission of culture from core to periphery. They note that innovation is usually assumed to emanate exclusively from the center, with peripheries being zones of imitation rather than innovation, but in actuality Lightfoot and Martinez point out that bi-directional exchange often creates a border zone charged with innovation and the resulting mix of traits is often the basis for the creation of local conceptions of identity. Thus, if we emphasize analysis from a local perspective rather than from one that assumes Ichmul's dependence, the different patterns of artifacts and settlement can be seen to be more the result of local opportunism than slavish imitation.

We must also consider the role history has in the reception of innovations by provincial centers. A simple model of colonial expansion is again insufficient. In most cases, larger centers rose to power in a settlement matrix already populated with lesser centers, as excavations have indicated was certainly the case with Ichmul de Morley. Lesser centers would themselves have acted and reacted to political and economic possibilities of the larger world in relation to local self-interest.

It appears here the border between Ek Balam and Chichén Itzá was not sharply demarcated and that Ichmul de Morley was not absorbed by either of its larger neighbors. If Chichén Itzá or Ek Balam controlled Ichmul de Morley, then most likely this was done hegemonically, and it may have been as much a local strategy as an imposition by its larger neighbors. Allegiances may also have shifted with rapidity impossible to resolve stratigraphically. And, of course, we must acknowledge that pottery and lithic distribution networks need not reflect political ties. To make further progress in determining ancient political territories, we need not only more regional surveys and excavations at smaller centers but also the development of bridging arguments that will help us understand the complex patterning of different classes of material culture.

Note

1. Batabilob and cuchcabalob were realms under the leadership of the *batab* (local leader) and the *ah cuchcab* (regional leader).

9

The Relationship between Tula and Chichén Itzá
Influences or Interactions?
Rafael Cobos

To some scholars, ancient central Mexico can be defined according to four cultural horizons. These cultural horizons are: the Olmec, corresponding to the Formative period; Teotihuacan, dating to the Early Classic period; the Toltec, dating to the latter part of the Classic period and initial part of the Postclassic period; and the Aztec, who dominated the Late Postclassic period and into the sixteenth century. These cultural horizons shared three basic characteristics: (1) they recognized a center of origin from which all influence arose and affected peripheral regions or faraway zones; (2) their influence was spread by people stemming from the center of origin; and (3) these people invaded, colonized, and eventually became the new rulers of the conquered territory or settlement.

An example of the aforementioned has also been documented in the Maya region between AD 700 and 1050/1100, particularly at the site of Chichén Itzá (see fig I.1). Since the late nineteenth century and early twentieth century, travelers such as Désiré Charnay (1887) and Edward Seler (1898) noted the presence of central Mexican features at that site and suggested that the Toltecs emigrated from Tula and were responsible for constructing the buildings of Chichén Itzá. From the 1930s until the end of the 1970s, scholars continued to argue that the Toltecs emigrated from Tula; however, they added two additional components to this argument. First, they believed that the Toltecs invaded the site, colonized it, and took over politically. Second, these scholars established a chronological framework for these events using their interpretations from ethnohistorical and historical documents from Central Mexico and Yucatán dating to the sixteenth and seventeenth centuries, as well as archaeological data. The chronological framework dated the end of the Classic Maya culture to AD 900 and the beginning of the Postclassic period in the tenth century, with Chichén Itzá as the most important site of the Maya culture during that period. This argument was fostered by the strong influence of the diffusionist model that predominated in anthropological and archaeological interpretation between 1930 and 1970.

The model of "migration, invasion, and conquest" continues to be favored by some investigators to explain cultural development and events that occurred in Chichén Itzá toward the end of the Classic period (Coe 1999; Diehl 1981, 1983, 1993; Proskouriakoff 1970; Wren and Schmidt 1990). The continued support of this model has much to do with the influence of the scholars who initially proposed it between the 1950s and 1970s. Alfred Tozzer (1957) and J. Eric S. Thompson (1970) were two prominent figures in Maya archaeology who dominated Maya studies and faced few challenges to their ideas (for a discussion of Eric Thompson's domination, see Coe 1999:123–44).

Nonetheless, Kubler (1961) recognized a weakness in the migration, invasion, and conquest model. For Kubler, the influence was unidirectional—stemming from Chichén Itzá to Tula—and the model did not recognize the possibility that it could have occurred the other way around. In a broader perspective, a similar argument was made in the 1960s and 1970s by archaeologists who felt that Teotihuacan held a great influence over various Maya sites during the Early Classic period, but who did not recognize a possible Maya influence at Teotihuacan (Braswell 2003; Cowgill 2003; Marcus 2003). Today, in light of new data recovered from Tula and Chichén Itzá, the relations that existed between these two important cities are better understood as "interactions" rather than "influences." The "influence" argued in the migration, invasion, and conquest model implies that the Maya were passive receptors of the culture and social mandates of Tula, while the interaction sphere model allows for the mutual exchange of cultural features and ideas.

The goal of this chapter is to demonstrate how the migration, invasion, and conquest model has dominated archaeological interpretation to explain the relationship between Tula and Chichén Itzá. Furthermore, I will evaluate a second model, which I refer to as the "local development" model, that has developed over the last decade and is currently favored to explain the relationship that existed between the great capitals of central Mexico and central Yucatán. In the local development model, interaction plays a key role in the interpretation of such relationships.

Previous Research

The Migration, Invasion, and Conquest Model

The migration, invasion, and conquest model applied to Chichén Itzá is widely based on ethnohistorical data, architectural features, certain sculptural representations, and iconographic elements. Some archaeological data are also considered in this model, and the data are tied to a specific ethnic group. At Chichén Itzá, these archaeological elements are usually associated with the Toltecs, an ethnic group that is believed to have emigrated from central Mexico, invaded the Yucatán, and conquered Chichén Itzá.

A traditional model argues that Chichén Itzá was founded and occupied by the Maya during the Classic period and later invaded by non-Maya individuals (Toltecs) or Nahuaticized Maya (Itzás or Putuns) who were responsible for the site's apogee during the tenth and eleventh centuries. This idea was first proposed by Charnay (1887), who used post-conquest documents to put together the legend tying the Toltec and Maya cultures. The legend states that a Toltec king by the name of Topiltzin-Quetzalcoatl was defeated by Tezcatlipoca and forced to leave Tula, Hidalgo. He is believed to have migrated with his followers to the east toward the Gulf Coast (Tabasco) and then on to the Yucatán (Coe 1999:167; Schele and Mathews 1998:356). Topiltzin-Quetzalcoatl and his followers arrived at Chichén Itzá and conquered it, establishing their cultural traditions and coexisting with the indigenous Maya population (A. Andrews and Robles Castellanos 1985; Coe 1999:167; Dutton 1952; Robles Castellanos and Andrews 1986; Thompson 1970: 3–47; Tozzer 1957). This model was further solidified when scholars tied the Topiltzin-Quetzalcoatl figure of the Toltec legend with the Kukulkan invader known from de Landa's *Relación* (Schele and Mathews 1998:199).

Alberto Ruz L'huillier (1964:209) further elaborated on the model by arguing that the Toltecs dominated the Maya and established a "Toltec pattern" of architecture, sculpture, ceramics, deities, and rites at Chichén Itzá. To corroborate his argument, he provided a detailed list of eleven architectural features, ten sculptural features, four ceramic elements, five deities, and three rites that Tula and Chichén Itzá shared.

One of the best-known examples of the "Toltec pattern" is evident in "the similarities that exist between the Temple of the Warriors complex from Chichén Itzá and the Temple of Tlahuizcalpantecuhtly, or Building B from Tula" (Maldonado and Kurjack 1993:100). These are structures formed by stepped platforms topped with multiple pillars that likely supported perishable roofs. However, at the Temple of the

Warriors, a wall on its summit surrounds the area where the pillars are located, whereas there is no wall on Building B at Tula.

The iconographic elements are dominated by reliefs of a plumed serpent (Quetzalcoatl), eagles and jaguars devouring hearts, processions of jaguars or pumas, and emblems of three arrows crossed (Maldonado and Kurjack 1993:100; Ruz L'huillier 1964:218). Sculptural representations are distinguished by warriors with shields, warriors located on pillars and door jambs, *chacmools*, Atlantean figures, standard bearers, serpent columns, and serpent heads set into balustrades (Ruz L'huillier 1964:218). Ruz L'huillier (1964:218) further argues that the presence of Silho Fine Orange ceramics, Tohil Plumbate, and green obsidian from Pachuca reflect Toltec control and distribution of these resources at Chichén Itzá. The list of Toltec traits is rounded out by the presence of cremation, the phallic cult, human heart sacrifice, and images of deities such as Venus, Tlaloc, Tezcatlipoca, Tlalchitonatiuh, and Quetzalcoatl-Kukulkan (Ruz L'huillier 1964:218).

More recently, Kelley (1992:116) has argued "Toltec Chichén and Maya Chichén (or Old Chichén) were constructed essentially during the same period, with Mixcoatl and Quetzalcoatl as rulers, dominating the center of the site, and Kakupacal as a venerated but subordinate Maya colleague." With this explanation, Kelley suggests that "Old Chichén" is associated with Kakupacal while "New Chichén" is associated with Kukulkan (see also Lincoln 1990).

Kelley bases his argument on the previously noted legends tying the two centers together through Quetzalcoatl-Kukulkan's arrival and on a new correlation that he proposes between the Maya and Christian calendars. Kelley (1992:113) believes that the arrival of Quetzalcoatl-Kukulkan was an actual historic event and that this event took place during Tozzer's (1957:40–43) Chichén III period (AD 1150–1260). Kelley's correlation differs radically from the Goodman-Martínez-Thompson correlation traditionally accepted and generally used in the Maya area by most scholars and proposes that "all the Maya dates [are] approximately 216 years after Thompson's correlation." Using this argument, Kelley (1992:118) dates the Osario or Structure 3C1 to AD 1214 and affirms that this structure is "a poor copy of the Castillo."

Diehl (1981, 1983, 1993) also argues that the apogee of Chichén Itzá during the tenth and eleventh centuries coincides, among other things, with the rise of the Toltec horizon represented by Tula, a "dominant polity and possibly the largest city of Mesoamerica." He sees the Toltec occupation of Chichén Itzá as part of an empire "of unknown dimensions, complexity, and duration" that the Toltecs established all over Mesoamerica (Diehl 1993:286).

In summary, the migration, invasion, and conquest model continues to be followed without critique of the legends stemming from post-contact documents. The "proof" that those events actually occurred are the new forms of architectural, sculptural, and iconographic elements at Chichén Itzá believed to have been established by the victorious group over the conquered one. In other words, these elements are seen as Toltec ethnic identifiers that were introduced to the site through aggressive action and victory in battle.

However, in light of new data uncovered over the last two decades, we must examine various arguments of the migration, invasion, and conquest model: first, whether or not the events documented in the historical sources occurred exactly as they are described; second, whether the architecture, sculpture, ceramics, deities, and rites actually represent the "Toltec pattern" at Chichén Itzá; and third, whether or not it is true that the non-Maya groups or Nahuaticized Maya present at the end of the Classic period were engaged in the use of force and warfare. A closer examination of these questions may support a new model that focuses instead on a local development. In this model, archaeological data and anthropological interpretations are utilized to demonstrate that the new ideas and styles that developed at Chichén Itzá reflect "interaction" rather than domination and that "there was more communication and less actual influx of new populations into the Maya country" (Jones 1995:76).

The Local Development Model in the Tula–Chichén Itzá Relationship

Despite the dominance of the migration, invasion, and conquest model over several decades, scholars such as Brinton (1882), Kubler (1961), and Proskouriakoff (1950) used historical documents, architecture, sculpture, and iconographic data to challenge that model. With the passage of time and the discovery of new archaeological data, along with new interpretations, the proposal of unidirectional influence from Tula to Chichén Itzá has slowly been weakened. Today, the combined result of this work makes it impossible to support the idea of a Toltec empire in Chichén Itzá and the rest of the Maya area (see Diehl 1993:286).

More recently, scholars such as Cobos (1997, 2001, 2003a, 2003b, 2004), Cobos and Winemiller (2001), Lincoln (1986, 1990), and R. Smith (1971) have used data from settlement patterns, ancient causeway systems, and ceramics from Chichén Itzá and have recognized Maya cultural attributes from the Late and Terminal Classic periods at that site. Furthermore, the study of all known hieroglyphic evidence supports the argument that the ancient inhabitants of Chichén Itzá were Yucatec-Maya speakers (Justeson et al. 1985; Krochock 1988; Ringle 1990), rather than speakers of a Mexicanized dialect or a central Mexico language. According to Maldonado and Kurjack (1993:100; see also Kurjack 1992), Chichén Itzá "did not constitute a Mexican site in Yucatán," nor was it conquered militarily by the Toltecs. Instead, they feel that the architectural, sculptural, and iconographic styles at Chichén Itzá reflect *contact* between the elite of Chichén Itzá, Tula, El Tajín, and other regions of Mesoamerica (Maldonado and Kurjack 1993).

For example, gallery-patio structures or patio structures without a frontal gallery have been reported at Tula and Chichén Itzá. Patios without frontal galleries appear to have originated at Monte Negro in Oaxaca during the Middle Formative period (Acosta and Romero 1992). In the eighth century, patios without frontal galleries were also in use at Alta Vista and La Quemada (Zacatecas) and at Chichén Itzá. Patio structures 5D3 and 3D11 (El Mercado) at Chichén Itzá appear to have integrated their frontal galleries in the tenth century (Cobos 1998, 2001, 2003a, 2003b). Patios without frontal galleries were found at Nohmul in Belize (D. Chase and Chase 1982) and Tula (Stocker and Healan 1989) and date to the ninth and tenth centuries, respectively. The archaeological evidence suggests that a patio structure without its frontal gallery was an earlier architectural innovation that was widespread in Mesoamerica and not simply a Toltec innovation forced onto Chichén Itzá.

A number of investigators have also observed the striking similarities between the Temple of the Warriors of Chichén Itzá and Building B of Tula. However, we should be cautious with these comparisons as Building B was completely reconstructed by Jorge Acosta during the last century. It has been noted that Building B "had been considerably destroyed" and that Acosta did not have information about what the structure looked like originally (Molina Montes 1982:130). Instead, Acosta employed data from Building C at Tula as well as the Temple of the Warriors at Chichén Itzá to reconstruct Building B. Furthermore, while Acosta (1945) reported to have found "48 imprints of pillars" in the building at Tula, "there has not been a single indication of the bodies of the pillars themselves. It seems that during a certain epoch everything was destroyed and the materials carried away" (Acosta 1945:48). Molina Montes (1982:131) also indicates that Acosta modeled the pillars in front of Building B after those used in the gallery from the Burned Palace at Tula. Molina Montes (1982:132) concludes that Building B from Tula is a product of reconstruction and falsification from the twentieth century and cannot be used in comparisons between Tula and Chichén Itzá.

Many of the pillars at Tula and Chichén Itzá include representations of individuals armed with darts, spear-throwers, and curved, grooved wooden sticks (Kristan-Graham 1992:11; Kurjack 1992; see also Coggins and Ladd 1992:244). Tozzer (1957) identified these individuals as Toltecs and noted that the Chichén-Toltecs also carry a small rounded shield on their backs, wear head ornaments decorated with birds, and have chest ornaments in the form of butterflies or birds (Kurjack 1992). According to Taube (1994:239), features such as the small rounded shield and feathered serpents identified by Tozzer as Toltec "can easily be traced back to earlier traditions of highland Mexico." He further notes that the iconography of the Chichén-Toltec "reveals a profound understanding and appreciation of ancient Maya belief, as well as traditions of the Gulf Coast and the Cotzumalhuapan region of Guatemala." In the iconography of Chichén Itzá, only the turquoise regalia appear to be an element of Toltec invention (Taube 1994:239), while the rest of the Chichén-Toltec iconography "suggests a self-conscious synthesis of Maya and Toltec traditions. Rather than being entirely eclipsed by Toltec influence, Maya traditions are clearly evident in all the themes that have been discussed" (Taube 1994:244).

Tozzer (1957) noted that on some of the pillars associated with the Temple of the Warriors there is a representation of a Maya defeat by the Toltecs. However, on other pillars and in other locations at Chichén Itzá, Tozzer discusses the contradictory images of Toltecs fighting among themselves and capturing each other (Kurjack 1992). Similarly, the iconography from six panels from the Great Ballcourt at Chichén Itzá has been interpreted as another space where a supposed Maya and Toltec skirmish can be appreciated. Based on the attire and objects in the panels, Tozzer (1957) identified separate Maya and Toltec ballgame teams. According to Tozzer, four of the panels represent Toltec victories, while the other two illustrate Maya victories. It should be noted that "the two Toltec defeats, however, are prominently illustrated at the center of the court" (Kurjack 1992:89). Kurjack (1992) has noted that Tozzer himself had problems understanding this spatial arrangement in the iconography of the Great Ballcourt since he believed the six panels should have exclusively depicted Toltec triumphs.

Another iconographic element not exclusive to the Toltecs is Quetzalcoatl. Representations of feathered serpents are widespread and exist from the Early Classic period at Teotihuacan. The Quetzalcoatl myth/cult, according to Ringle et al. (1998), may have been initiated in the sixth century, and revitalized during the tenth and eleventh centuries. Gillespie (1989:123–207) argues that the belief in this deity could even postdate European contact in Mesoamerica.

In a recent work, Cobos (2000) argues that Quetzalcoatl, or Kukulkan, was an important sovereign from Chichén Itzá who ruled the site sometime during the tenth century. The *Relaciones histórico-geográficas de la gobernación de Yucatán* (*Historical-Geographical Relations of the Yucatán Government*) (de la Garza 1983) registers the name of Kukulkan as a great lord and paramount ruler of Chichén Itzá, and more recently, Lincoln (1990) and Cobos (2000) recognized that Captain Serpent and Kukulkan were the same individual present at that settlement during the tenth century. This may be supported by the iconography on the North Temple of the Great Ballcourt and the Temple of the Chacmool, as it portrays the enthronement of Captain Serpent, or Kukulkan. We do not know the origin of this important sovereign known at Chichén Itzá and later at Mayapán as Kukulkan. However, considering the time between the collapse of Chichén Itzá in the eleventh century (A. Andrews et al. 2003; Cobos 2004) and the rise of Mayapán in the thirteenth century, it could not be the same individual that founded Mayapán and gave the name to this settlement as de Landa (1959:13) documented. Rather, it appears that after the death of the ruler named Kukulkan at Chichén Itzá, later sovereigns that governed Chichén Itzá and Mayapán might have used the term Kukulkan as a title associated with the highest political authority.

The migration, invasion, and conquest model proposes that the Toltecs were in charge of distributing Pachuca obsidian, Silho Fine Orange ceramics, and Tohil Plumbate to Chichén Itzá and the rest of Mesoamerica. Some scholars, however, have documented that the distribution of green obsidian in the Maya area dates to the Early Classic period (Braswell 2003), centuries before the emergence of the Tollan Phase (AD 900–1150) at Tula (Mastache et al. 2002). Although Pachuca obsidian found at Chichén Itzá accounts for 18 percent of the site's collection, it is not the only source from which obsidian was obtained (Braswell 1997:plate 1). According to Braswell (1997), Chichén Itzá is the only site known in all of Mesoamerica that obtained obsidian from ten different sources. This contrasts dramatically with the obsidian procurement of Tula during the Tollan Phase where "about 90 percent of the city's obsidian is green material from Sierra de Pachuca, Hidalgo" (Mastache et al. 2002:43; see also Healan 1993). The sharp difference in supply patterns for obsidian at Tula and Chichén Itzá seem to demonstrate that the Toltecs were not in charge of controlling the flow of obsidian toward Chichén Itzá.

Silho Fine Orange ceramics have been found in stratigraphic contexts dating to the middle of the eighth century at Isla Cerritos (Cobos 1997:22) and during the ninth and tenth centuries at other sites from the northern Maya lowlands (Cobos 2004:522). Chichén Itzá controlled the distribution of Silho Fine Orange in the Maya lowlands, although this pottery appears not to have reached Tula or not to have been exchanged between Chichén Itzá and Tula (Cobean and Mastache 1987; Mastache et al. 2002:46–50). On the other hand, Cobean (1990; see also Mastache et al. 2002:46) recognizes other orange and cream ceramic types in Tula and suggests that they are related with traditions from the Gulf Coast of Mexico, although the precise place of origin has not been determined.

The center of origin of Tohil Plumbate has been located in the coastal plain of western Guatemala, and it was initially produced during the second half of the ninth century, almost half a century before the Tollan Phase at Tula (Cobos 2004:522; Neff 2002). Tohil Plumbate is defined as a commercial pottery that was exported from its center of production to different regions of Mesoamerica, northern Mesoamerica, and Central America and reached various ethnic groups. Mastache et al. (2002:48) argue that Tula "probably controlled its commerce, its internal distribution, and its redistribution to other regions" of Mesoamerica. I differ from the point of view expressed by Mastache et al. (2002:48) about the exclusivity that Tula had over the distribution and commercial control of Tohil Plumbate in all of Mesoamerica, although I do recognize that Tula could have been the site that controlled the regional distribution of Tohil Plumbate in northern Mesoamerica.

An important group of braziers and *incensarios* have been found at Tula, much like those known at Balankanche Cave near Chichén Itzá. The braziers from Tula correspond to the type Abra brown-coarse and show six varieties. Included among these are the plain variety in the form of an hourglass with the image of Tlaloc, the cylinder form variety, and the hourglass variety with appliqué (Diehl 1993:280–81, fig. 10; Mastache et al. 2002:48, fig. 3.1). At Balankanche Cave, these incensarios correspond to the Chichén unslipped ware and show similar forms to those reported at Tula (E. Andrews IV 1970:plates 2, 8–13). The similarities between the braziers and incensarios found at Tula and Balankanche Cave made Cobean (1990:508) think that Maya artists created the ceramics under the direction of priests or artists from Tula. Mastache et al. (2002:48), however, argue that the braziers from Tula had diverse origins from those at Chichén Itzá. Furthermore, braziers with attributes similar to those reported at Tula and Balankanche Cave are present in the Maya lowlands from the Formative period. This suggests that the Toltecs from the Tollan Phase did not create nor export braziers of the type Abra brown-coarse toward Chichén Itzá.

Archaeological analysis of ceramics from Tula and Chichén Itzá indicate independent local developments as well as ones from regions located short distances from both cities. For example, in the case of the Tula pottery, the Tollan ceramic complex defined by Cobean includes approximately twenty-five ceramic types whose origins can be found in different traditions (Mastache et al. 2002:46–50; see also Cobean 1990; Cobean and Mastache 1987). These traditions include the red-on-brown ceramic from central Mexico; orange and cream from the Gulf Coast; Huastec regional ceramics; Mixtec incensarios (Alicia Openworked); Blanco Levantado type from the Bajío-Guanajuato region; and Tohil Plumbate from the Maya area.

At Chichén Itzá, the Sotuta ceramic complex dominates and includes Chichen Unslipped, Chichen Slate, Chichen Red, Fine Orange, and Tohil Plumbate (R. Smith 1971:134–35). Of these, the most prevalent ceramic wares at the site include Chichen Unslipped, Chichen Slate, and Chichen Red, originating from central and western Yucatán and southern Campeche. Tohil Plumbate and Silho Fine Orange ceramics originated in the coastal plain from western Guatemala and the lower Usumacinta region, respectively, and are represented in the Chichén Itzá collection in smaller numbers in comparison to the majority wares. Ceramicists George Brainerd (1958) and Robert Smith (1971), with more than a quarter of a century of ceramic studies in the northern Maya lowlands, were in charge of analyzing the ceramics recovered by the Carnegie Institute of Washington in more than ten years of archaeological research at Chichén Itzá. After extensive analysis, they concluded that there was a local ceramic tradition, with continuity between Chichén Itzá and other settlements in Yucatán. As Robert Smith (1971:253) stated, "Sotuta pottery was locally made by Maya potters with relatively little influence from Tula or Mexico."

Discussion and Conclusions

Throughout this chapter I have demonstrated how the same group of archaeological data has been interpreted with completely different points of view. In part, the migration, invasion, and conquest model emerged as an explanation of the events and cultural processes in Mesoamerica that historical sources noted for the final part of the Classic and Epiclassic periods and the beginning of the Postclassic period. In this model, the archaeological data are examined and fitted to the historical explanation. On the other hand, the local development model utilizes archaeological and historical data that have to be compared with archaeological interpretation.

As was demonstrated above, the arguments that favor one model over the other have produced radically opposing conclusions. While the first model indicates that there were migrations of individuals, the second suggests instead that there was intensive communication between different regions. The architectural, sculptural, and iconographic elements employed by the first model to prove migration, invasion, and conquest are refuted by the second model, which argues that some of these elements have their origins in the Early Classic period. Furthermore, some of these features, such as ceramics, appear to be local to Chichén Itzá or the southern Maya lowlands. The migration, invasion, and conquest model associates certain archaeological features with an ethnic identity that forced its elements onto the site of Chichén Itzá through violence or domination. However, many of these features were widely distributed in Mesoamerica and could have been seen as commodities or objects of prestige that were acquired by different elites and/or ethnic groups.

In conclusion, the local development model favors the communication between regions and suggests that societies with different social and cultural trajectories could have interacted at any moment during their development. These interactions could have been manifested at the elite level, and it is here where we must focus our attention and efforts in research. For example, understanding how these interactions occurred, distinguishing the different moments in which these interactions were manifested, and determining the detail of different levels of intensity between them still represents a challenge in the archaeological investigation. Future archaeological research that continues to analyze interactions—instead of influences—at the end of the Classic and Epiclassic periods between central Mexico and central Yucatán have an enormous investigative potential. These studies are still in their infancy and works such as this one may help them to reach maturity.

Note

In memoriam
Alba Guadalupe Mastache

Part 4 Today's Scene

10

Ethnoarchaeology in the Northern Maya Lowlands

A Case Study at Naranjal, Quintana Roo

Kurt R. Heidelberg and Dominique Rissolo

In the humid tropics, we often find that residential and agricultural spaces overlap. The convergence of home, garden, workspace, and orchard confuses the landscape, as the absence of any clear delimitation precludes the separation of functional areas that are more obvious in other regions. Such is the case among the lowland Maya of southern Mexico and Central America. Living in these garden residences, the modern Maya are at the center of a complex environment, with components such as living quarters, workshops, gardens, orchards, refuse dumps, and animal pens.

Work has been done to determine spatial components of these types of residences throughout the Western Hemisphere (Covich and Nickerson 1966; Denevan and Schwerin 1978; Hiraoka 1986; Johnson 1983; Johnston and Gonlin 1998; Killion 1990, 1992; Kimber 1973; Robin 2002; Vogt 1969). We find a wealth of literature on the modern lowland Maya homegarden, most often concentrating on distributions of plant species (e.g., E. Anderson 1993; Caballero 1988; Herrera Castro 1994; Ortega 1993). Some effort has been made to apply the ethnoarchaeological value of such studies to the interpretation of ancient sites in the same area (see de Pierrebourg 1999).

Research Problems

In the state of Veracruz, Thomas Killion has developed a model for characterizing houselots, the Household Garden–Residence Association (HGRA) model (for all references to the HGRA model below, see Killion 1992). Though not originally intended for the Maya area, this model, through ethnographic analogy, shows much promise for applicability in the northern Maya lowlands.

A difficulty with ethnographic analogy is coming to terms with changes in ideologies, politics, religion, climate, and assorted other factors that control the way in which people live. This is particularly so in the Yucatán, which saw drastic changes in politics, population, and available species at the introduction of the Spanish and in later tumultuous times. This chapter is in two parts. First, it will evaluate the applicability of Killion's model to modern houselots in the Maya town of Naranjal in Quintana Roo, Mexico. Next, it will consider some anticipated variability in the model due to historical changes that have occurred over the last five hundred years.

In his model, Killion defined five major areas within each houselot: a "Structural Core" for sleeping, cooking and eating; a "Clear Area," which serves as a location for diversified activities as well as a staging zone for trips and activities to be continued elsewhere; an "Intermediate Area," defined by an accumulation of materials swept from the Clear Area; a "Garden/Orchard Area"; and a "Lot Boundary." In anticipation that this model would apply well to the northern Maya lowlands, the senior author chose to evaluate its applicability there. Such an effort might help us in our understanding of domestic space use among the ancient Maya.

The Research Study

The community chosen for the study, Naranjal, is a small village of approximately fourteen families in northeastern Quintana Roo. Their means of subsistence is largely based on maize agriculture, and income is supplemented through the sale of honey, handicrafts, charcoal, some cattle, and potted ornamental plants that some families raise in their houselots. Most of these products are sold to businesses in Cancún, which is approximately 85 km away. The families raise nutritional supplements within their houselots, or *solares*, in the form of chickens, turkeys, pigs, and assorted fruit and vegetable crops. The houselots are packed tightly together, with many shared walls. Outfield agriculture is practiced in the outer periphery of the *ejido* (common land) property.

Methods

This study was part of a larger research project intended to inventory types and proximity of economic plant species in Yucatec Maya solares. Information was collected to determine covariation among plant species and specific activity areas. In the course of data collection, detailed mapping was performed, recording size and location of all structures and features including: animal pens, agricultural features, hearths, cooking pits, and fences, as well as more generally defined loci such as toilet areas and trash-burning areas. Care was taken to record construction features, materials used, doorways, windows, and paths. All plants within the solares were identified and plotted, and the head of the household was interviewed to determine the origin of the plant (weed, cultivated, or deliberately planted) and its uses, if any. The head of the household was also questioned on the use of specific areas, whether delineated by fences, structures, vegetation, or variant management, as was often the case in Killion's investigations.

For the sake of this study, we observed the spatial distribution of physical components and activities in most of the solares, some in great detail. It was the senior author's intention to develop a model of association between specific activity areas and patterns of waste deposition within the houselot. It was apparent that Killion's model maps remarkably well onto the domestic landscape at Naranjal, and though the characteristics of these areas vary some from the characteristics of those in Killion's model, we chose to adopt most of this terminology for the present study. Though the nature of the model remains the same, some of the specifics vary between the two regions. Following is a discussion of each of these areas in turn, with their characteristics, in the village of Naranjal. One finds that our descriptions of the village landscape bear a strong resemblance to those found in earlier authoritative ethnographic works (e.g., Redfield and Villa Rojas 1934). Nevertheless, in order to highlight the applicability of Killion's model in the northern Maya lowlands as well as to evaluate its potential role in future ethnoarchaeological research, such descriptions may prove valuable.

Structural Core

The Structural Core in houselots at Naranjal is composed of one or more apsidal or rectangular structures in which food preparation, eating, and sleeping take place. These are nearly always clustered in close proximity to one another. Although it is not unusual for a family to have more than one dormitory structure, it is rare that any structure is used solely for food preparation. Those in which food preparation take place are usually treated as general living spaces, and with a three-stone hearth in a front corner, the remainder of space typically contains hammocks, family memorabilia, and often a shrine. They are also used for storage of food and recyclables inside and larger items under the roof dripline outside. These structures are typically constructed of poles for walls, and roofs of *guano* palm leaves or corrugated tarpaper. Some of the structures have slab floors and most have skirts of stone around their perimeter to keep debris from blowing in through the spaces between the poles. A few families in Naranjal have dormitory structures built of concrete blocks. All are normally situated in a high spot within the houselot, often on outcroppings of limestone, and stabilized with a rubble and earth platform. With few exceptions, the Structural Core lies very close to the edge of the houselot that provides the primary means of access to the property.

Clear Area

The Clear Area surrounds the Structural Core and is where a large portion of domestic activity takes place. Diversified activities take place here, such that it serves as a general family living area and kitchen when the weather permits. Leisure activities and reception of guests take place here, as well as parties. When at home, children tend to stay within this area where the older members of their families can supervise them. The Clear Area serves as a work space for temporary projects and as a staging zone to prepare for trips to tend bees or livestock, or to outfield areas for agriculture. Food items are often processed for storage here, such as the drying of *chiles*.

The Clear Area is kept well compacted and free from debris through trampling and daily sweeping. In the Clear Area directly adjacent to the Structural Core, one often finds elevated planting beds (*ka'ancheob*) and recycled cans, pails, and tubs used for raising medicinal and cooking herbs, as well as seedlings that may later be transported to the *milpa* (corn field). These features are also found at the edges of the Clear Area, adjacent to the Intermediate Area. This proximity allows for constant monitoring from the Structural Core, sufficient sunlight for such plants, and protection from animals. This also permits Structural Core site selection strategies to ignore direct availability of soils, since the planting surfaces are portable. Many-coursed stone rings (*wo'ol tunich*) to protect larger herb or vegetable plants are also found throughout the Clear Area. Bathing and dressing structures are also often located in the Clear Area.

Intermediate Area

Debris swept from the Clear Area is deposited into the bordering Intermediate Area. The level of yard maintenance drops significantly in this zone, as does the level of any daily human activity. Rocks, plastic, broken ceramics, and assorted other bits of garbage litter this region. Household waste is frequently dumped here and rapidly decomposes or is eaten by animals. Often, inorganic waste such as plastic is kept in a single location, sometimes contained by a pen or walled area. Compost formed from decay of organic waste frequently provides volunteer cucurbits, chiles, and tomatoes, which are sometimes protected and cultivated. Old soils from the ka'ancheob kept in the Clear Area are dumped here. Occasionally, debris is raked into piles and burned. Weeds are cut as needed, but economical wild species are maintained and non-indigenous trees, such as orange and lime, are not rare.

Activities that traditionally take place in the Clear Area may sometimes spill over into the Intermediate Area, particularly large projects that generate much waste or cooking for parties and festivals. One often finds extra hearths here, as well as a favored spot for pit cooking.

Animal pens fall on the break between the Clear Area and the Intermediate Area. These pens keep the animals far enough from the Structural Core to reduce the impact of any associated odors on the living space and keep the Clear Area free from unnecessary obstructions. The animal structures are nearly always within view from the doorways of the buildings in the Structural Core, allowing for constant monitoring with minimal effort. The pens also act as a barricade behind which large garbage items and recyclables can be hidden or stored. They are constructed of saplings, guano palm, and corrugated tarpaper. Latrine facilities are often found in walled sections within the Intermediate Area. Additionally, storage structures are sometimes found in the Intermediate Area.

Although neighbors often do not have walls separating their properties, in none of the observed instances did Intermediate Areas from two solares stand adjacent to each other. Even in cases where brothers lived in adjacent lots, their Clear Areas were treated as an individual unit, with no Intermediate Area in between. In the event that neighbors do have substantial walls of stone, the Intermediate Area normally will fully occupy the space between the Clear Area and the wall.

Garden/Orchard Area

There is an extraordinary amount of variation among the gardens in Naranjal. For the sake of the model, the Garden/Orchard Area is found beyond the Intermediate Area and does not include the ka'ancheob and other small planting surfaces found within the Clear Area. This area is better described as an orchard, as it consists primarily of fruit trees and cultivated wild species. Trees to provide wood for construction material or firewood, such as tropical cedar, are usually maintained among the fruit species. Though infrequent, some have small monocultural plots in which cucurbits, maize, or sorghum is raised.

The higher-story vegetation in the Garden/Orchard Area blocks out sunlight well enough that the floor is relatively free of weeds. Exceptions to this are normally found in clusters of young citrus trees, which must be kept clear of weeds through occasional cutting. The Garden/Orchard Area extends toward the back of the property, bounded by the Lot Boundary or an unmanaged area. This area serves as a natural wall, which does not necessarily define the property bounds of the houselot. Densities of waste materials will generally increase at the edge of such unmanaged areas. Some families intentionally leave a section of this brush to serve as a privacy screen containing an area for latrine use.

Lot Boundary

The Lot Boundary is normally well defined, with stone walls, barbed wire, hedges of bushes, and small trees, and in Naranjal, remains of ancient structural mounds. This boundary does not always define the used area of the houselot but rather the legal bounds of the property. In many cases, neighbors leave an unused area between lots when space is not at a premium. This allows for more privacy than the alternative wall systems. Many stone walls found in communities throughout the northern Maya lowlands are actually ancient *albarradas*, which have been left intact or altered to accommodate more recent property divisions or spatial requirements.

Waste Management at the Homes in Naranjal

As in other areas, the site formation processes are not so simple in Naranjal that we should depend on walls, foundations, hearths, or other such discrete features as our only determinants for archaeological interpretation. These features are short-lived and are made of organic materials and of stone that may be carried off for some other project when they are no longer of value for a current purpose.

By observing domestic space use through a model such as the one presented, we weigh more importance on the waste-depositional trends associated with use and, more importantly, maintenance of space. Clearly, the location where a broken pot is ultimately deposited is rarely the same as that place where it was broken. By observing waste management among spaces in conjunction with space use, we find regular relationships among the defined areas across all of the solares in the study, such that the location of waste alone is sufficient to reconstruct the general layout and orientation of the site. Though the majority of human activity takes place in the Clear Area, it is here that we find the least amount of debris. Constant sweeping away from the Clear Area results in a band rich in waste material, defining the Intermediate Area as well as leaving characteristic deposits along the edge of the Lot Boundary. Beyond the Intermediate Area, we find a gradual decrease in debris, relative to distance from the Clear Area. As this space becomes the Garden/Orchard Area, we find more evenly disbursed debris resulting from less structured depositional patterns. Finally, along the edges of the Lot Boundary we may find an increase in debris resulting from efforts to keep the Garden/Orchard Area clear.

Post-Contact Effects: Conquest and Assimilation

By the middle of the sixteenth century, the Spanish exercised considerable control over the lifeways of the Yucatec Maya. Implementation of an *encomienda* system (the system by which conquistadors in Mexico were guaranteed indigenous labor and tribute to accompany lands granted by the Spanish crown) brought new demands on Maya subsistence practices, as now they were required to furnish foreign goods to the Spanish elite (see Patch 1993:28–32). Though many of these tribute goods were indigenous—such as maize, honey, turkey, cotton, beans, and wax—demands for recently acquired Spanish products clearly existed. There is no doubt that houselot agriculture was common among the Classic and Postclassic Maya (Dunning and Beach 2000; Lohse and Findlay 2000; Pyburn 1998; Robin 2002). Introduction of new species to the area was extraordinary. Prior to European contact, the only domesticated animals to be found within the northern Maya houselot were the oscillated turkey, Muscovy duck, and deer. Chickens and hogs, for example, worked their way into the Maya household almost as quickly as the arrival of the Spanish themselves, becoming a common component only a few years after conquest (Clendinnen 1987:31). Introduction of Spanish fruits and vegetables came quickly, too. Indeed, Bernal Díaz del Castillo, when chronicling his journey with Córtez in the conquest of Mexico, claimed to have been the first to sow orange seeds in the region decades before the Yucatán was taken (Díaz del Castillo 1956).

The need to furnish extra food as tribute was not necessarily new: such practices had been in place prior to the introduction of the Spanish. Regardless, with new players on the receiving end of the tribute, this practice must have affected activities and priorities in the houselot. However, the effect of raising a few extra crops quite likely pales in comparison to the simple presence of the Spaniards: new people with a new culture bringing new products and practices to the region. If pigs and chickens were not raised as tribute items, it is reasonable to assume that they still would have worked their way quickly into Maya food-ways. Both animal and plant species were exchanged between Maya and Spanish to the extent that over 25 percent of the species in modern Maya solares are European in origin (Barrera Marin 1980). This seems to be a global phenomenon witnessed during the meeting of two food-ways. When comparing dooryard gardens in southeast Asia to those in southeast Mexico, E. Anderson (1993) noticed that both regions have borrowed species from each other over the course of time and that the gardens are now remarkably similar in terms of borrowed species. This is not to suggest a lack of diversity among houselots across the region, however. In modern surveys, floristic variation in houselots across the peninsula remains substantial (Caballero 1988).

It is clear that foods raised in the solar have changed drastically in the last five hundred years. What is not as clear is the effect that these changes have had on archaeological signatures, how they fit in the HGRA model, and how ultimately they might be interpreted. From a general standpoint, little has changed: food was being raised. What we should anticipate, however, is that changing the relative impor-tance of species may affect the layout and use of the houselot. For ex-ample, prior to contact, most tree crops could be found in the wild. Do-mesticated ducks and turkeys could roam freely about, requiring little space of their own. The introduction of citrus fruit trees, pigs, cows, and horses changed this balance. Unlike the case with pre-contact tree species, anyone desiring citrus could not choose to harvest in the for-est. In modern times, horses, cattle, and pigs are often penned within the houselot. These shifts in prioritization of plant and animal species have had an effect on solar space management. To accommodate these new products, the homeowner had two options: increase the size of the solar or make sacrifices with the existing space. The first option is not very practical, particularly when communities are already established and houselots share boundary walls with one another. Indeed, the trend seems to have been a reduction in the size of solares since con-tact. During the Spanish-imposed congregation of Maya communities in the sixteenth century, the people were inserted into a more urban environment where houselots were deliberately made to be smaller and generally the same size (Farriss 1984:159). This trend of scaling down appears to continue into modern times (Flores Peña 1993).

The second option—a change in the management of domestic space

—was the more practical one. Changes in houselot structure could happen within the individual "areas" in the solar. For example, introduction of citrus could result simply in a sacrifice of other trees in the Garden/Orchard Area, such that the general management of the area remained the same. Introduction of new, and likely more, livestock posed a greater threat of encroachment from one space to another as well as affecting the placement of those spaces. Corrals for horses or cattle tend to be placed within the Garden/Orchard Area. Pens for hogs and chickens, as well as new support structures, create a "thickening" of the boundary between the Clear Area and the Intermediate Area. The placement of these pens further guides the order of the Structural Core such that the increased livestock activity can be monitored from doorways.

Introduction of new technologies by the Spanish have had their effects on yard management. Perhaps a more conspicuous example is the introduction of steel and subsequently the machete. Lithic debitage was managed in earlier times and is virtually non-existent in the modern household (Vanden Bosch 1999; Walling et al. 1999). Subsequently, deciding how stone tool working affects the characteristics of the model is problematic. Assorted strategies for management of lithic waste could have been in place: stone work could have been performed in the Clear Area with the debitage swept into the Intermediate Area; special areas may have been set aside for knapping, farther away from the Structural Core where the sharp flakes would be less of a danger. More important is how that practicality of performing certain tasks within certain areas changes with the shift from stone to steel. Requiring far less maintenance, a relatively large amount of certain types of work could be done with a machete, where if the work were done with stone tools, it would result in extraordinary amounts of debitage and a need to carry in large amounts of tool stone to the houselot. Large cutting tasks, previously performed outside of the houselot, may have moved within at the introduction of the machete. Some processing activities (butchering, construction preparation, etc.) that take place in the Clear Area today may well have been performed in the Intermediate Area or Garden/Orchard Area to reduce the danger and inconvenience associated with stone tool preparation and sharpening.

Changes in agriculture and animal husbandry suggest certain shifts within individual houselot sites, but shifts in inter-site structure came with the Spanish as well. Rules for social reorganization imposed by the Catholic Church changed the structure of Maya communities. Existing social hierarchies within communities were broken down to reduce the status of any indigenous people controlling organizational

structure. The Maya elites suddenly found themselves at the same social status as their subordinates. Efforts were made to congregate normally dispersed groups of Maya by reforming towns with a *zócalo* (town
square), church, and *audiencia* (town hall) as town center, with houselots surrounding the remaining circumference and extending outward
(Farriss 1984:158–59). The physical changes to houselot structure are
apparent in the implementation of these policies. However, the nature
of the spatial separation between houselot and milpa is more likely a
function of the region's unique physical geography (Restall 1997:169–
70).

It is equally important to consider the effect of social reorganization on the Mayas' interactions with each other, and how subsequent
changes in debt, hospitality, resource distribution, communication,
and transportation eventually have an effect on the nature of the
houselot. Congregation was a culturally destructive process that made
a moral and symbolic delineation between town and country, setting
up a dyadic code in which the newer communities were associated
with Christianity and civilization, and the forest was associated with
wild beasts and untamed forces (Farriss 1984). The effect that such a
philosophy must have had on Maya perception of domestic space and
the place of the houselot within its larger, natural environment must
have been phenomenal. Towns such as Naranjal, despite their small-
ness in size, reflect this recent delineation through the importance
placed on the zócalo as the "center" of town. Straight roads and contiguous houselots located around the square stand in stark contrast to
the adjacent forest. The very means of subsistence has changed to be
more conducive to a more regimented town environment and general modernization. Many Maya now work for money or raise or produce goods such as honey, charcoal, or potted plants for sale. These
industries are generally male-dominated, but the effect on the solar
is not necessarily negligible. Staging for these activities requires more
space when done on a grand scale, and we may witness more thickening of the Intermediate Area for storage of equipment. The potted
plants, for sale in Cancún, are raised in the Garden/Orchard Area closer
to the Structural Core, sometimes encroaching on the Intermediate
and Clear Areas to exploit the sunlight that tends to fall within these
spaces.

Discussion and Conclusions

In his study of modern Nunamiut campsites, Lewis Binford (1982) illustrated how, due to the simple nature of being human, we can antici-

pate certain elements in a site framework. People eat, sleep, and create garbage, and are limited by laws of physics, their size, and constraints of their physiology. Activities tend to be localized and performed with some degree of organization. Clearly, an indeterminable amount of change has affected domestic space use and management in the northern Maya lowlands, but these changes are not prohibitive when applying a model such as HGRA. Much in the spirit of Binford's rationale, we should still anticipate patterning at some general level. By capturing the nature of how space is maintained through signatures in waste management, we are capable of characterizing zones for classes of activities, if not the specific nature of the activities themselves. By relying on such a model, we may reconstruct spatial distribution of the discussed activity area, even in the total absence of architectural features. This is of particular importance in the Naranjal area, as Postclassic reoccupation of the major Formative site surrounding the modern village resulted in the robbing of wall and foundation materials from earlier solares to build later ones. Modern repopulation has had the same effect. These alterations to the landscape have little effect, however, on the waste that was left behind. Evaluating modern Maya houselots in juxtaposition with regional domestic archaeology gives a framework by which to discuss layout and variability. Our understanding of changes that have taken place in time and that have affected the use of space allows us to scrutinize inconsistencies between the modern and ancient signatures. Though such systems of quantification are not always rich in answers, they provide a vehicle by which we may ask questions.

We anticipate that excavation of, and the application of this model to, the waste materials within ancient houselots in the region will aid us in developing a clearer interpretation of the layout and, importantly, use of ancient Maya domestic space. Through observation of relative densities of that waste as discussed here, clearer viewpoints on the lifeways of the ancient residents can be constructed.

11

Archaeologists Working with the Contemporary Yucatec Maya

Dominique Rissolo and Jennifer P. Mathews

The nature of an archaeological project often requires that researchers establish a temporary residence in a local community. Concern for the conditions that affect, and are affected by, their presence in this new place and space is often considered peripheral to the task of realizing research objectives. In fact, many archaeologists would admit to enjoying a certain sense of security in their perceived temporal, and therefore legitimized, dislocation from their object of study. In the most extreme of cases, an archaeologist might resemble a geologist—extracting, observing, or examining symbolically inert physical material with little regard to contemporary cultural contexts.

Nevertheless, the discipline of archaeology has become increasingly more active in efforts to bridge the gap between the archaeological record (as it has been recovered and interpreted by archaeologists) and those peoples who claim it as part of their heritage (see Dongoske et al. 2000; Downum and Price 1999; Ford 1999; Lynott and Wylie 2000; Marshall 2002; Pokotylo and Guppy 1999; Sabloff 1998). However, "being" an archaeologist—that is, an "outsider"—can be as much at issue as "doing" archaeology when living among the people who inhabit the location that has been designated for study. Most ethnographers are trained, in some way, to deal with the challenges of living in the communities in which they work. Most archaeologists are not trained ethnographers or applied anthropologists. Yet, we often occupy, and even become part of, a community that is not our own. This chapter will examine some of the current literature and research related to archaeologists working with descendant communities, the ways in which a Maya community–based project differs from a non-community-based project in the northern lowlands, as well as provide examples of some of our positive and negative experiences in working with local peoples, in the hopes that other projects can benefit from this knowledge.

Current Research Problems and Questions

Nowhere else in the Maya area is the cultural and physical geography of an indigenous people as contested, by such a diversity of interests, as in the northern lowlands. The unparalleled combination of the dramatic Caribbean coastline, lush tropical forests, warm climate, stately haciendas, spectacular ancient ruins, and thriving Yucatec Maya communities continues to prove fertile ground for all manner of ventures—be they entrepreneurial, advocacy-based, or research-oriented. Cancún, one of the ten largest resort destinations in the world (Rider 1999:101), is the epicenter of this regional transformation (Re Cruz 1996b). Its presence is felt well beyond the tourist zone as *ejidatarios* (people who farm collectively owned lands) migrate to Cancún and other attractions along the "Riviera Maya" and across the peninsula in order to take advantage of wage-labor opportunities. Needless to say, the subsequent effects on the social and economic fabric of Maya village life are significant and have lately received considerable attention from scholars (see Brown 1999; Castañeda 1996; Hervik 1999a, 1999b).

A number of recent critiques explore the potential for indigenous or local descendant communities to benefit, in a variety of ways, from the *internal* management of their patrimony (Ardren 2002; Ardren et al. 2000; Griffith and Colwell 2000; Wille et al. 2000). The extent to which this alternative socioeconomic strategy involves archaeologists falls under the rubric of public, applied, or community archaeology and the emerging concepts of cultural tourism or heritage tourism. Such scholars have also taken the discipline of archaeology to task for its all-too-common indifference towards indigenous concerns, as well as its exclusive or partisan portrayal of the archaeological record. Fortunately, an increasing number of archaeologists share Ardren's "goal of practicing an archaeology that is both more responsive and more relevant to our host communities" (2002:396), while heeding Pyburn and Wilk's (2000) caution of the potential pitfalls, pratfalls, and paternalism of well-intentioned community archaeology.

K. Anne Pyburn (2003) calls for a more "engaged" Maya archaeology rather than the detached and ostensibly objective approach of traditional field research. She affirms that "[t]he issue of the social context of archaeology is not a stand against science; it is a desire for better science and more responsible interaction with the present" (2003:289). Mayanist scholarship stands only to benefit from an honest and informed approach to community archaeology, and we have little to lose by abandoning the obsolete template of "pure" scientific inquiry. We feel that our "engagement" should resemble the deference and flexibility of partnering, not the hubris and intransigence of parenting. Applying what the archaeologist thinks is best for the community is not applied archaeology; in fact, it misses the point entirely.

Other researchers have examined aspects of the contemporary Maya interacting with their ancient past—a past that is often (re)created by contemporary archaeologists and anthropologists. Peter Hervik (1999a:59–90) discusses the "external constructions of 'the Maya'" using three examples of outsiders' interpretations of Maya culture: Spanish invaders, writers and photographers of *National Geographic*, and writers in Western newspapers. Using ideas taken from cultural studies, he examines the concept that these types of representations report information that shapes the reader's understanding and perceptions of the subject. He argues that the subjects of these texts are not given a voice in these contexts but rather are described by outsiders for an audience of readers that are totally unrelated to the subject. On an associated topic, Victor Montejo (1997) has also strongly criticized archaeology's role in "world building" and sees it as inappropriate for Western scientists to interpret and "create" the archaeological past of the ancient Maya. Montejo feels that outsiders are imposing their perspectives and shaping the view of the Maya that they deem appropriate, rather than allowing the Maya to interpret their own past.

In his volume *In the Museum of Maya Culture: Touring Chichén Itzá*, author Quetzil Castañeda (1996) examines anthropology's role in the "commoditization of culture" in the northern lowlands. Using a postmodern approach, he argues that at Chichén Itzá, Maya culture has been continuously reinvented and has involved the complex interface between the local peoples, anthropology, and the development of tourism. He feels that this culture was never "pure" and unchanging but rather in a constant state of flux that cannot be defined with a single "truth" about the past (1996:16–20). While specific to ethnography, his discussion is clearly applicable to archaeology and our discussion here as well.

Yaeger and Borgstede (2003) examine the social history of the practice of Maya archaeology. They examine how early European and European-American exploration and scholarship created the "disjunction" between the splendor and mystery of an ancient civilization and the peasantry of its descendant population. That legacy has in many ways contributed to the perpetuation of the very real social, economic, and political divisions we witness today. Yaeger and Borgstede compel us to appreciate the complex mosaic of contemporary Maya identities and realities and therefore stress the need for "context-sensitive strategies" with respect to research (2003:277). Entering the field with

a monolithic notion of the modern Maya is naïve at best and dangerous at worst.

These researchers implore archaeologists and anthropologists to examine their methods and interpretations and the way in which they
circulate their information. It is in this spirit of self-examination and
reflection that we would like to share a page from our journal of experiences of living and working as archaeologists in the contemporary Maya community of Naranjal. We did not initially set out in our
research to negotiate issues of community interaction; nevertheless,
we quickly came to realize the urgency and ethical imperative of such
efforts. Fortunately, we are not alone in our desire to learn from, and
share with, other Mayanist archaeologists—from our host nations and
beyond—that are actively engaged in community-based projects. Only
through such an open forum (unbounded by ethnicity or nationality)
can the discipline of archaeology hope to remain viable and dynamic,
while community concerns are actively addressed.

"Being" an Archaeologist in a
Contemporary Maya Community:
Challenges Faced in a Community-Based Project

The village of Naranjal is located in northern Quintana Roo near the
Yucatán border and is further discussed in the chapter in this volume by Heidelberg and Rissolo on ethnoarchaeology. The community consists of fourteen households and controls an *ejido* (collective
land holding), which was established in the 1950s. Most residents enjoy what they consider to be a traditional way of life, as characterized
primarily by their language, religion, and what the men and women
describe as their responsibilities to their families and their community. Government-funded infrastructure improvements are present—
some more welcomed than others are—and Naranjal maintains contacts with other communities and agencies.

The community of Naranjal also happens to reside amidst the ruins
of an ancient Maya center known as El Naranjal. The unique attributes
of this major site captured the attention of the Yalahau Regional Human Ecology Project and so began our relationship with the resident
community. In 1993, we entered into our first field season at El Naranjal as commuters. Project directors Scott Fedick and Karl Taube negotiated a large-scale mapping effort that enlisted the men of the village
to help clear the structures. Our choice to live off-site was essentially a
practical one, but it also allowed both the archaeologists and the community to test the waters.

Our seasonal residency in the community began in 1996. Like many researchers, we returned to the field with both personal and professional agendas. After all, our primary objective was to complete successfully the tasks that we explicitly stated in our proposals. Though we were extended an official welcome by the community and were already involved in friendly, if not business-like, relationships with the men, we soon realized that our tenure in the community would force us to reevaluate the nature and purpose of our presence, as well as our roles and responsibilities as guests and agents of change.

During this first residential season, one of the first difficulties that we faced working with a community as archaeologists was that our archaeology permit issued by the Instituto Nacional de Antropología e Historia (INAH) did not necessarily guarantee the community's blessing. A few members of the village expressed some hesitation about our working on the ancient ruins, and we were asked not to begin work until the issue was resolved. Fortunately, after some discussion with the men about what we hoped to accomplish during our field season, they prescribed a course of action, which involved a contract between the people of Naranjal and the project. This was ultimately brokered and approved by the Consejo Supremo Maya—a non-governmental, grassroots, indigenous-rights group based in a nearby community. This involved us bringing the head of the organization to the village to oversee the meetings. Under his watch, it was also agreed that the community leadership would draft the work schedule and roster of workers and would set the wage as they had done in 1993. After the terms were agreed upon, we wrote up a formal contract, which was then signed by us and the community leaders.

Another issue that we faced was when the men of the village expressed concern about the handling of the archaeological material, which they considered to be ejido property. It seemed that they were not only concerned about what we were going to do with the materials but also what role INAH played in curation of the artifacts. In an effort to demystify our activities at the site as well as the function of INAH in this process, we offered to take a few of the men to the INAH office in Cancún, where they could visit the museum and the facility that handles much of the archaeological material recovered in northern Quintana Roo. They held a meeting to discuss the offer and selected ten men for the trip. Sara Novelo, the Director of the INAH Museum at that time, gave the men a tour of the facilities, discussed the conservation methods for archaeological remains, and explained that any artifacts brought to the museum from El Naranjal would be returned to the community after being conserved.

Fig. 11.1 The women of Naranjal gather at the basketball court for a Mother's Day celebration (Photo by Dominique Rissolo, 1996.)

In an effort to establish a connection and bring the museum to the community, Ms. Novelo and two other INAH employees made special arrangements to transport a Chac effigy censer to Naranjal and display it in one of the town's public buildings. Throughout the day, village residents wandered in to talk with Ms. Novelo or simply to take a look at the figure. It was never our intention to elicit a particular response as a result of these interactions. Such impressions—be they positive, negative, or ambivalent—are personal and private and become part of the community's individual and collective experience. Additionally, tangible efforts to inform the community of the details of our work include the dissemination of our proposals, reports, and maps (in Spanish), as well as periodic public discussions of our progress (fig. 11.1). In our experience, the more transparent we were about our efforts, the less often misunderstandings or conflicts with members of the community arose.

As would be expected, living in Naranjal as archaeologists and as outsiders has forced us to deal with complex issues of interaction and intervention. Many archaeologists might express a sincere and often misguided concern regarding their impact on village life. However, according to Castañeda, "[t]he idea of impact implicitly and explicitly argues that the society or culture being impacted is a static, ahistori-

cal, agencyless, solidly bounded, non-interactive object, whether conceptualized (imagined by social scientists) as an organism, a system, a
structure, or a text" (1996:9). We would add that it is arrogant to assume that indigenous communities do not have the social structures
to accommodate our presence. Moreover, it would be foolish to under-
estimate their ability to manage our activities effectively.

In return for the privilege to live and work in Naranjal, we are obligated to participate in certain aspects of village life, as deemed appropriate or necessary by the men in the community. Below are descriptions of a few incidents that have occurred over the last several years
at Naranjal that exemplify how living in a Maya community has affected both our relationships with people in the community and our
archaeological research.

A specific event that made clear our fundamental role in our professional relationship with the community involved the vandalizing
of a well-preserved ancient structure at the site. The damage was significant enough to warrant an assessment by INAH, and the act itself,
which was believed to have been perpetrated by young men from a
nearby community, was seen as both a violation of ejido sovereignty
and the destruction of communal property. We surveyed the damage
with the men and offered our opinions when solicited, but we were
hesitant to intervene. We came to realize, however, that intervention
was never an issue. The course of action was decided by the men during general meetings, as it always has been. Our role was to carry out a
specific task that was delegated by the ejido leadership. In this case, we
were asked to contact the appropriate agencies and authorities and ar-
range a meeting at the site, since Naranjal lacks radio communication,
telephone, or regular car service. If we had not been present in Naranjal at the time of the incident, they would have no doubt pursued the
same course of action by different means. Regardless of our concerns
about involvement in community affairs, the ejido leadership is more
than able to distinguish between intervention and implementation.

Our function as a conduit to the outside also involves late-night
trips to the nearest doctor or purchasing hard-to-find medicines in
Cancún or Valladolid. Though we are often obligated to assist, most
people in the community do not approach such requests casually. It is
our understanding that they take pride in their self-reliance and will
only approach us if the task is seen as a personal favor or if all other possibilities for assistance have been exhausted. Either way, people in the
community always make a gesture of payment for our services, which
is politely refused.

Each field season coincides with a number of holidays and celebra-

Fig. 11.2 The crosses from the Catholic church in Naranjal are removed for a Day of the Holy Cross procession. (Photo by Dominique Rissolo, 2000.)

tions, such as Children's Day, graduation, or the Day of the Holy Cross (fig. 11.2). We are always welcomed as guests and are expected to make modest contributions of candy, food items, or the local alcoholic beverage of *aguardiente* (distilled cane liquor). Such events put archaeological fieldwork on hold—often for several days—requiring us to adjust schedules and staffing. To continue working during the celebrations, in Naranjal or elsewhere, might not necessarily be prohibited but would certainly be considered rude and inappropriate. Additionally, participating in these events can be a wonderful experience that may allow archaeologists to strengthen their ties to the community and draw a greater understanding of the contemporary cultural traditions.

Although our being welcome in Naranjal is partly contingent upon our ability to provide work for the local men, our relationship with the community is not strictly professional. The process of living and working together has formed bonds of friendship and *confianza* (mutual trust), as well as relationships more ambivalent in nature. In 1997, we were asked to sponsor three children (from a family we had grown close to) for their First Holy Communion. Becoming *compadres* (co-parents) is not something we had planned on, but we felt honored and accepted the offer and the responsibilities that such a position entails. We cannot speculate on how this was interpreted by other families in the community, nor could we ever expect a consensus regarding people's feelings toward us as individuals and as outsiders. Nonetheless, it is our feeling that when taking on a responsibility such as this, it is important to understand the implications of this kind of relationship and to act appropriately. These kinds of social ties to members of the village may imply relationships that could outlast the time spent conducting research there and require a commitment to return simply out of social obligations and friendship.

We, of course, are not the only outsiders that interact with the community, but we are, if not generally trusted, at least a known quantity. The people of Naranjal are well aware of the value of archaeology in terms of attracting tourists and generating revenue and are already making preparations to accommodate the controlled trickle of tour groups that are starting to arrive. Some of the men have begun carving small wooden figures or images such as the Maize God, which they sell to small groups of visiting tourists. Additionally, several women in the community are producing handicrafts on a larger scale, such as *huipiles* (embroidered dresses), embroidered *servilletas* (cloth napkins used as tortilla warmers), and hammocks. The capacity in which we are to participate in the development of tourism at the site is dependent upon numerous factors, which have not yet been fully discussed. However, as outsiders—like the tourists—our assessment of current activities at the site was often considered.

When a film crew expressed interest in shooting part of a pseudo-documentary in Naranjal, the ejido leadership met, weighed the pros and cons, settled on a price, and granted them permission. We were subsequently notified of the date and were welcomed to observe. Although the entire production appeared to us as contrived and as a harbinger of unwholesome changes to come, life in a Maya community cannot be expected to fulfill the archaeologist's personal notions of purity and authenticity. We have come to recognize that the community, in their own way, will dictate their level of interaction with outsiders and decide what influences are permissible.

Toward Establishing a Collaborative Research Environment

Developments ranging from private so-called "ecotourism" parks to more traditional large-scale corporate tourism ventures have attempted to integrate ancient and contemporary Maya culture into their attractions. Such ventures have been successful in terms of drawing tourists but offer nothing meaningful in the way of interpretive or didactic programs and essentially rely upon financial might or political maneuvering to gain access to local patrimony. Alternative and more substantive cultural experiences available to the visitors of the northern lowlands include archaeological sites and museums managed by INAH, federally recognized biosphere preserves, and small-scale, ejido-based cultural tourism or development efforts. It is in this latter category that "outsider" archaeologists often find themselves. The ejido is also the contemporary spatial and social framework in which many archaeologists or ethnographers conduct their research into past and present Maya culture in the northern lowlands. These researchers are becoming more involved in the collaborative preservation and development of cultural resources (which range from archaeological ruins to indigenous knowledge) into both viable sources of communal revenue and incentives for Maya individuals to remain part of ejido life.

Unfortunately, our fascination with the Maya of the northern lowlands, however well intentioned, has often led to the appropriation and commoditization of indigenous patrimony for the advancement of external economic, and even academic, capital. A step toward reversing this trend is to engage communities in the design, implementation, and monitoring of all activities involving cultural resources (in a manner that is permissible by local and federal regulations). For the Maya of the northern lowlands, the ejido remains the fundamental essentializing component of community identity and cultural affiliation —regardless of whether or not ejido sovereignty is imagined or real. Therefore, efforts to facilitate community involvement in this process must begin at the ejido level.

It is clear that local *ejidatarios* are both fully aware of their potential as active agents in the process of local and regional development and possess the level of sophistication necessary to negotiate their collective destiny strategically. They are familiar with the push-and-pull factors created by international interest in their cultural patrimony and have dynamic social structures in place to address such challenges. The real issue is whether or not ejidos are sufficiently empowered to mitigate, control, or direct the nature and degree of external interest in the archaeology and cultural resources of the northern lowlands.

A major impediment in this regard can be the archaeologists' indifference towards creating a collaborative environment in which roles and responsibilities of both sides can be openly discussed. An additional obstacle for local communities is the lack of reference tools (i.e., archaeological literature) needed to make informed decisions and to articulate ejido interests in community-based research effectively. It is not for the archaeologist to dictate or impose a management strategy in this regard, but we have the unique opportunity to reevaluate the information we generate for both U.S. and Mexican institutional sectors and make that information available to those individuals it directly and indirectly affects. As Pyburn and Wilk stress, it is our responsibility to make the archaeological record "intellectually accessible" (2000:79). The dissemination of our intellectual property (which is essentially the transliteration of ancient and living indigenous knowledge) is an integral part of our relationships with local communities. It is also our responsibility as archaeologists to maintain the balance of confianza while acknowledging the ejido's authority to determine the course of its interactions with outsiders.

Discussion and Conclusions

Each year we return to our respective study areas, not only as archaeologists but also as positive or negative reminders of what we have done and what we have the potential to do. "Being" archaeologists and "doing" archaeology constantly requires us to reconcile personal and professional expectations with those of the community. There is no doubt that our choice to conduct archaeological projects in local communities (on the communities' own terms) means that we accept the possibility of altered research designs, prolonged field studies, controlled access, or awkward negotiations. It may also mean that we enjoy the support of some members of the village, while others would prefer that our research was not carried out there. In some extreme cases, rumors or mistrust may trigger leaders within a community to decide not to allow archaeological research to be conducted at all, resulting in the project having to relocate entirely. However, in these instances, the situation may be entirely dependent upon the opinions of the current leadership and could change with subsequent elections.

As archaeologists living and working in the northern lowlands, we are privileged to be offered a unique and invaluable perspective into Maya culture. In our view, we feel that living and working with contemporary Maya peoples necessitates not only the responsibility of understanding our roles in the village, but the realization that the villages will change and progress whether we are there or not. We hope that as resident archaeologists we will be able to offer resources and information that members of the community can take into consideration when facing challenges or making decisions affecting the future of their collective patrimony. With respect to our long-term commitments to the people and places that have given to us the vital core of our field research, the things we do as archaeologists are not checks on some life-list of academic experiences but are more akin to signatures on a contract—the terms of which are not our own.

In this chapter, we have outlined some of the experiences we have had living in a contemporary Maya village in the hopes that other researchers in similar environs can benefit from our knowledge. It is our expectation that archaeologists in the northern lowlands will discuss more openly the challenges that confront them in communities like Naranjal, and that we will not begin our pursuit of past human behavior without first carefully examining our own.

12

Mílpas of Corn and Tourism Mílpas

Alícia Re Cruz

Much has been said about the repercussions of the act of tourism, primarily in the form of "packages" to popular international tourist destinations. Mexico greatly satisfies the growing global demand for ecological and culturally oriented holidays through the promotion of Cancún as the most acclaimed tourist emporium. While in Cancún, tourists breathe in Maya culture when they go on tours to nearby archaeological sites and when they experience the cultural bits and pieces spread throughout the tourist market (i.e., hotels resembling Maya pyramids, monumental Maya sculptures decorating entrances of commercial malls, or waitresses wearing *huipiles* [traditional female Maya dresses]). Through out-migration, Maya peasant communities across the Yucatán Peninsula provide the labor force that supports the tourist industry and urban development of Cancún. These Maya migrants become the underlying base of labor for the tourist industry and for development.

Tourism breaks down cultural borders by bringing together "hosts" and "guests" as well as their cultural baggage. In the process, images and cultural representations are created and reinterpreted within the socioeconomic and ideological postulates of the tourism rationale. In Cancún, thousands of individual experiences are interwoven: tourists encounter the exoticism of a Maya ancient culture that is kept alive by the exhibition of the archaeological remnants; at the same time, Maya migrants seek their dreams of progress and self-improvement. The Maya culture that migrants bring to Cancún is quite different from the Maya cultural paraphernalia manipulated for tourist marketing purposes. However, throughout their experiences as low-wage laborers, upon which the complex capitalist tourism apparatus rests, migrants in Cancún learn the crucial role that Maya cultural "authenticity" plays in the economic success of the tourism enterprise.

I have written extensively on the role of tourism in the lives of Yucatec Maya communities (see Re Cruz 1996a, 1996b, 2003). The latest ethnographic chapter on the connections between Cancún tourism, migration, and contemporary Maya communities is a recent ethnographic documentary—*Los Otros: Maya Migrants in Cancún*—a project that required several short field visits to Chan Kom, the Maya community discussed in this chapter, and Cancún from 1997 to 2003. Most of the recent ethnographic data presented in this chapter emerges from this, my latest ethnographic production, which introduces an important element in the discussion of the repercussions of Cancún in the current transformations of Maya culture: how do the Maya negotiate and manipulate the concept of "authenticity of Maya culture" to be consumed in the tourism market?

Previous Scholarship

A fruitful venue of criticism in current anthropological analysis stresses the relations of power in which peoples and places become "the object" of the tourist's desire. The publication of *Orientalism* (Said 1979) has paved the way for the elaboration of these serious critiques against international tourism as the epistemological force that either converts peoples and their cultures into "commodities" (Greenwood 1989; Munt 1994) or imposes neo-colonial networks of power upon the local communities (Britton 1982; Nash 1989). The scholarly literature has also charged tourism with ridding local cultures of their historical roots, pointing out a crucial issue in the study of tourism—that of "cultural authenticity" (Greenwood 1989), or how cultures in their past and present are being constructed under the gaze of tourism (Urry 1990). Through our anthropological lens, we are attempting to understand how cultures and communities are "imagined" (B. Anderson 1983). In the tourism context, there are quite a few agents involved in the process of "imagining" culture: the tourists who want to live and experience "the exoticism"; the government boasting about the resources of the country (Lanfant 1995:33); and the local people promoting their cultural ancestry in order to be recognized as members of the tourist experience. Rather than continuing the research venue of ratifying the social and cultural heresies committed in local communities in the name of tourism, this chapter focuses on how locals deconstruct their experiences with tourism as migrants and apply these lessons in "imagining" their local communities within the tourism market. This chapter also engages the debate on "cultural authenticity" by applying this research strategy to Chan Kom, a Maya community with a long tradition in dealing with "guests"; it explores the dexterity that the Maya exhibit in interweaving cultural constructs of their ancient past with current representations and forces emanating from the Maya experiences within Cancún tourism.

Chan Kom, the Ethnographic Face
of the Tourist Experience

More than half of the Maya in Chan Kom have experience as migrants
in Cancún (see Re Cruz 1996a). These migrants have to cross not only
the rural/urban border but also different cultural and socioeconomic
borders as they move between living and working in Cancún and Chan
Kom. In their trips back and forth, they bring their cultural framework
and experiences with them. Out-migration from Chan Kom has become a crucial factor in the sociopolitical fragmentation of the community (Re Cruz 1996a). The Maya presence in the proletariat barrios
of Cancún has strongly contributed to the mosaic of different ethnic
groups, social classes, and nationalities that color the cultural landscape of Cancún's tourism (Re Cruz 1996b). In their endeavors to embrace the capitalist economy, the migrant Maya have "imagined" a
Mayanized Cancún, conceptualizing the city as their new *milpa* (corn-
field)(Re Cruz 2003). Meanwhile, in their response against the influences of Cancún, Maya peasants who remain in Chan Kom are "imagining" Cancún as a dangerous place ruled by money, in which people
are assaulted in their homes, murdered, and raped (see Re Cruz 2003).

Cancún's tourism has led to social fragmentation within the village. Since the late 1980s, a group of Maya from a leading family in
the community has been monopolizing the political *cargos* (positions)
of Chan Kom. These individuals are brothers with long experiences as
migrants in Cancún. Throughout their histories as Cancún laborers,
some of them have become business entrepreneurs, owners of bakeries
and fruit stores, and contractors. Through interviews conducted during a brief visit to Chan Kom in 2001, I began to hear about their plans
to encourage tourism development in the community:

> I see the municipality as having a future—a future lies ahead of it. What we have
> right now is that we are at the point of inaugurating the road that is 9 km from
> the intersection of Mérida and Valladolid. This will benefit the peasant sector.
> The road was built thirty years ago. It was made by hand, so it had a lot of im-
> perfections. So, with this, the producers from the south will bring their prod-
> ucts from there to the city of Cancún. In the old time, peasant production was
> carried out by horses; right now the peasants get their harvest out, and right
> now we have the ability to take it to Cancún to sell. And Chan Kom, in about ten
> years, will become a very important place. Touristically speaking because we are
> very close to Chichén Itzá, which is about fifteen minutes away from here and
> we have the airport of K'aua about fifteen minutes away. So all of that is very
> beneficial for us and Chan Kom. Chan Kom will be a very prosperous munici-

pality with a lot of jobs. . . . We have right here a *cenote* (karstic sinkhole) that in the old times people used to take water out to drink, and in that cenote a tunnel has been discovered so we hope it can become another tourist attraction for Chan Kom. . . . So I see for Chan Kom a very promising future so our children can obtain their much-desired dreams—and I hope that we can see it happen, the dream to move Chan Kom ahead. (Beto Cime, Chan Kom's president, personal communication 2001)[1]

However, before further discussing the social fragmentation that has developed in the village, let me first provide you with a brief ethnographic history of Chan Kom so that we can better understand how the migrant experience in Cancún tourism is inspiring local plans to fulfill "the dream to move Chan Kom ahead" by offering the community cultural capital to serve the tourist market.

Chan Kom, a Historical and Ethnographic Portrait

Chan Kom is located at the heart of the north-central part of the Yucatán Peninsula (see fig. I.1). Two major archaeological sites flank the community: Chichén Itzá (14 km north) and Cobá (90 km east). The community is 9 km off the Mérida-Cancún highway, 132 km from Mérida and 290 km from Cancún. The name "Chan Kom" derives from the existence of the central cenote (*chan* in Yucatec means "little" and *kom* means "kettle"). The entire Yucatán Peninsula is a karstic plain; the rainwater filters through the rock, eroding it and creating caves and cenotes (also called *dzonot* in Maya) or sinkholes (see Houck, this vol.). According to the Maya Yucatec oral tradition, cenotes are believed to be the entrances to the underworld; they are the entrance to an intriguing cave and underground river system. These cenotes served as refuge areas for the Maya throughout the colonial times and modern historical events such as the Caste War and the Mexican Revolution. An abundance of pre-Hispanic ritual paraphernalia has been found in most of these cenotes, and today they are still imbued with a feeling of "sacredness."

Other remnants from the pre-Hispanic past of Chan Kom are visible on the stones from masonry buildings and in the low mounds dispersed around the milpas; the Maya call these mounds *muulo'ob* (small hills), believed to be the houses of the *aluxo'ob* (lords of the milpa). It is not uncommon for Maya peasants to find little clay or stone figurines or other pre-Hispanic remnants in these muulo'ob. These archaeological remains are seen as manifestations of the glorious past, as well as

reminders of the cyclical nature of events so ingrained in Maya episte-
mology.

According to Robert Redfield and Alfonso Villa Rojas (1934), the first
inhabitants of Chan Kom were descendants of the Maya who took the
side of the Yucatec government during the Caste War of 1848. The first
peasants came from Ebtun, 45 km away, looking for fertile land to cultivate milpas. The peasant displacement provoked by the Mexican Revolution (1910–20) also affected the Yucatán Peninsula, and this initial
settlement also attracted displaced peasants from other communities.
After the revolution, Chan Kom received the first legal grant of *ejido*
property (collectively-owned village land), and it was assigned *pueblo*
status, as a legal and political entity independent from Ebtun. Finally,
it became a *municipio libre* (free township), and in 1935 Chan Kom became the *cabecera* (center) of a group of peripheral *comisarias* (hamlets).
Chan Kom started with 251 inhabitants in 1930 (Redfield and Villa
Rojas 1934:13); it reached 530 in 1972 (Elmendorf and Merrill 1977,
1978) and increased to 682 in 1990 (Re Cruz 1996a).

What has been written about Chan Kom represents a valuable entry
within the encyclopedic knowledge of Yucatec Maya culture. This
knowledge starts in 1924 when the Carnegie Institution began a
twenty-year investigation in and around Chichén Itzá. Numerous
modern Maya villages surrounded these magnificent Maya ruins uncovered by archaeologists. The American anthropologist Robert Redfield turned to these Maya communities to find ethnographic information not only to interpret the past but also to understand the role
of that past in modern Maya communities. When Chan Kom became
the focus of ethnographic attention, it was at this point that its residents had the first contact with "guests" or foreigners. Soon, following
the Carnegie Institution and Redfield's ethnographic project, the first
Protestant missionaries arrived in the area.

Alfonso Villa Rojas, born and raised in Mérida, was a schoolteacher
in Chan Kom. He was recruited by "el Americano" (i.e., Redfield) as an
assistant in the field research of Chan Kom (Sullivan 1989:33). Redfield and Villa Rojas (1934) initiated the ethnographic investigation
by looking at the contemporary Maya of Chan Kom as the model of a
"peasant society" within the "folk-urban continuum," a model to conceptualize cultural change (Redfield 1941). Within this model, Chan
Kom represented the peasant stage of development and Redfield imagined a socially harmonious, family-oriented community, still attached
to a rather romantic past. Preoccupied with the need to document
change and development, Redfield returned to Chan Kom to write the
insightful *A Village That Chose Progress: Chan Kom Revisited* (1950). Red-

field identified that the progress in the village was being propelled by a combination of the Maya's own work and technical assistance from the government and from foreigners like the Americans at Chichén Itzá (Sullivan 1989:158). Despite the indications of development and progress that Redfield found in Chan Kom, he was harshly criticized by Victor Goldkind (1965, 1966). While under the Redfieldian umbrella of homogeneous "peasant" society, Goldkind discovered social schisms that resulted in a Protestant exodus of Maya from Chan Kom to Pisté, the nearest community to the archaeological site of Chichén Itzá. The presence of foreigners, primarily the Americans in Chichén Itzá and Pisté, not only activated the spread of Protestantism in Chan Kom but also acted as a magnetic pull for Chan Kom's Protestants and their allies.

Chan Kom's acquaintance with anthropological research continued in the 1970s with Mary Elmendorf's (1972, 1976; see also Elmendorf and Merrill 1977, 1978) studies on women's productive and reproductive roles in the community life. It was the first time that the focus of the study of Maya peasant society was on women's economic and social contributions. The most recent chapter in Chan Kom's ethnographic encyclopedia analyzed the impact of out-migration to Cancún on the socioeconomic, political, and cultural dimensions of community life (Re Cruz 1996a, 1996b). These different chapters of anthropological analysis of Chan Kom have promoted the popularity of this community among those familiar not only with Maya culture but also with community studies, cultural change, women's studies (production and reproduction), migration, and tourism, among many other anthropological foci of inquiry. Visiting Chan Kom has become a ritual pilgrimage among anthropology apprentices and professionals, as well as for tourists who, on the way to Chichén Itzá, venture to Chan Kom in order to feel "the experience with the living Maya." Via anthropological research, tourism, or a combination of both, Chan Kom's contact with "guests" has become ingrained in its history.

Confluence of Current Maya Identities in Chan Kom

The Protestant missionaries who came to Chan Kom around the 1920s contributed to the expansion of Protestantism in the community, which paralleled the growth of cattle development (Goldkind 1965, 1966). The old enmity between two of the founding families (Cime and Pat) was religiously expressed in the confrontation between Catholics and Protestants. This religious division was also tied to an emerging source of wealth (cattle ownership) as a major threat and possible political competitor for the Cime, the family of the *cacique* (political leader). The confrontation between the Cime family and the emerging competitors (members of the cattle-owning Pat family) is remembered as a battle between Catholics and Protestants. The socioreligious clash ended with the Protestant exodus from Chan Kom in the 1950s. The defeat of the Pats allowed the assertion of the monopoly of political power by the Cimes.

The initial migration of young people to Cancún in the early 1970s opened new social fissures in the community. Some of the young members of the Cime family were among this first migrant group, and they accumulated enough capital to build their own enterprises in Cancún. Increasing enrichment of these individuals and their families did not go unnoticed by peasant leaders who were threatened by the competition and influx of cash. During the 1987 political election, two PRI political candidates were appointed. One was a young owner of a fruit store and a bakery in Cancún, elected by the group of migrants. The other was a schoolteacher promoted by the peasants who had remained in the village. The migrant's victory in the political election aggravated a social schism and started a serious social irony in the community: the political leadership was in the hands of individuals "on the move," conducting their business in Cancún, while still performing their duties in Chan Kom.

Community Division and Debate on Maya Identity

The Maya defeated by the migrant's leadership became known as *los Antiguos*, alluding to their connection with the ancestors, and with it, a strong determinant of their Maya identity. This term is often used as a synonym of *milpero* (worker of the milpa). Within the Maya cultural logic, the milpa defines the peasant's identity as Maya. The milpa is the system of Maya relationships wherein the individual who produces corn, the sociocultural Maya order, and the corn itself are intimately linked. Through corn production, the milpero maintains a harmonious relationship with the cosmos and the sacred world; it is through *primicias*[2] that Maya peasants are granted divine permission to work the lands of Nature. Corn is brought up from Nature by male hands and once harvested is "civilized" and transformed into *tortillas* by female hands. For the Maya, tortillas are *la fuerza* (the strength), and the energy that moves people to work and to live. Once it is planted, corn is conceived as following the Maya life cycle—from its birth to its maturity and death in the fields—and as having an afterlife of nurturing the Maya through the tortillas produced by Maya women.

According to the *Popol Vuh*, corn is the flesh and bone of the Fifth World people, the most perfect humankind. This perfection emerges from the harmonious connection among the cosmos, nature, and the Maya. Thus, corn symbolically encapsulates the cosmic order in perfect harmony and balance between natural and metaphysical forces. It also reaffirms the cyclical nature of cosmic, natural, and earthly events. It is through corn production that humans attempt to replicate cosmic harmony on earth. Collectively, this is realized through adherence to rituals. Therefore, corn and the ritual paraphernalia associated with it are symbolic representations of "harmonious balance" among the three alter-egos of the Maya cultural logic: cosmos, nature, and body (see Re Cruz 1996b).

Los Antiguos legitimize their Maya identity by calling themselves milperos because they continue to be attached, economically and culturally, to milpa work. With this, they not only reaffirm their Maya identity by situating the milpa as their epistemological and productive foundation, but they also attempt to de-identify the migrant group as Maya. Making migrants "non-Maya" disqualifies them as legitimate power holders in community affairs. However, as intermediaries of geographical settings or cultural brokers between the community and Cancún, Chan Kom migrants transfer their familiar cultural code to the new urban setting. Maya migrants from Chan Kom, particularly those who migrated in the early 1970s, described Cancún as a *milperío* (field house). In fact, most of the Maya communities have their origins as a milperío, which eventually develops into a village. Furthermore, Maya migrants continue identifying themselves as Maya, explaining that their milpas are not in Chan Kom anymore, but in Cancún. These new milpas are the restaurants, hotels, and wages that are the result of a much different relationship between producer and product. Cultivation of corn links present and past in a continuing cycle; anytime a Maya person is engaged in milpa production, he or she repeats the actions of those first creators of the most perfect humankind possible, the Corn People. In the urban milpas, product (wages and money) and producer (migrant worker) belong to different cultural rationales (see Re Cruz 2003). Furthermore, Maya migrants get used to the idea, and consequently learn, that Maya tradition can be transformed into a tangible "thing." With their savings, migrants prefer to pay other milperos in the village to work their own lands while they work in Cancún. Paying others detaches them from the exercise of the peasant traditional knowledge, which connects the individual with nature and the cosmos via ritual performance. Thus, traditional peasant knowledge is translated into an object, represented by the Maya culture consumed in Cancún tourism. It follows that Cancún, ontologically speaking, does not fall within the cycles of time and events according to the Maya paradigm; it does not have past or future because it does not have a connection that links it to the Maya ancestors. This is why it is not uncommon to hear about predictions that Cancún will be tragically destroyed. This is illustrated by the following description from a Maya migrant from Chan Kom:

> It is said that all Cancún is going to be flooded, not even cars will be able to escape. Everyone is going to drown, until no one is left there, even the hotels . . . everything will disappear. It is said that the *Yuntziles*[3] are going to close the four corners of Cancún and they are going to sink it. . . . 'Kan' in Maya means serpent and 'Kun' means that it is like a big jar so that is the name of Cancún, the *X-Kukikan*[4] that dwells in the sea that is going to turn over, and it is going to send a great rain, and everyone is going to die. So my wife is very scared, and she wants to go back to Chan Kom. (don Alito, personal communication, Cancún 1990)

Tourism and Culture in Chan Kom

Members of the PRI party, who dominated political leadership in Mexico until 2002, have monopolized the political cargos in Chan Kom since 1987. Certainly, PRI support has been a crucial factor in the solidification and strengthening of the migrant leaders' power in community affairs. The national-community power connection is nurtured by a series of negotiated exchanges from which los Antiguos feel excluded. In reality, the community leaders become a very effective tool for the government to entice Maya into the "dream" to prosper in Cancún. In fact, the members of current political leadership are examples that demonstrate that through work, migrants can "make it" and become rich. This process is tremendously gratifying for the nation, since the tourist industry is the main economic revenue for the country. Thus, channeling migrants towards Cancún is a national service that current Chan Kom leaders brilliantly utilize in exchange for PRI political support. Exhibitions of this type of political nepotism in community affairs has since moved los Antiguos leadership to turn toward PAN, the major political opponent to the PRI party.

While in Cancún serving the tourist industry, migrants become familiar with the idea that Maya culture can be sold, negotiated, advertised, promoted, and packaged. Thus, they learn the keys to market Maya culture for the tourist industry. Furthermore, the current community's political leaders are also familiar with the historical interest that Maya archaeologists and anthropologists have shown for Maya culture and for their community in particular. During my 1989–90 field research in Chan Kom, I asked the president of Chan Kom, the owner of a few capitalist enterprises in Cancún and a member of the Cime family, about the origins of the community. He went inside his house and came out a few minutes later with a copy of *Chan Kom: A Maya Village* (1934) to indicate that the book could provide all the information I requested. Furthermore, he insisted that the book was about his grandfather don Eustaquio Cime, the founder of the village as Redfield specified in the book, which includes don Eus's autobiography. It was an outstanding move in legitimizing his power as the community's political leader, providing a rebuttal against los Antiguos' accusations of holding the office illegitimately because he is a migrant who abandoned milpa work and rituals a long time ago.

It seems that there is a connection between nation, state, tourism, and communities. In an effort to mobilize the cultural heritage for tourist consumption, the nation seeks to raise public consciousness regarding the need to exhibit cultural memorabilia. It follows that tourism unwittingly can result in the promotion of national political interests. In Chan Kom's case, the community becomes a replica of the national efforts to bolster cultural aesthetic consumption. Maya migrants in Cancún learn how tourism can be seduced by nationally sponsored versions of Maya history and culture. Mayas are also aware of the positive value attached to the "Maya" label when they witness and experience the tourists' interest in ancient Maya culture and their demand for goods that reflect that culture. Thus, along with the tourist exploitation of cenotes and caves with ritual paraphernalia in Chan Kom, the current president plans to open artisan stores, a small hotel for those tourists who decide to spend the night in the village, and restaurants offering handmade tortillas and other Maya fare for the tourist palate.

Discussion and Conclusions

In exploring the malleability of the Maya cultural rationale, this chapter has addressed Maya agency in responding to the commoditization of culture in the tourist enterprise, sometimes interpreted as a neocolonial form of exploitation. However, this analysis has pursued a different avenue primarily focused on the efforts to react against tourism or to use the elements of the tourist-market rationale to be implemented in the community. Ritual and everyday practices associated with milpa production yield an ontological power enabling los Antiguos to engage in an identity fight against those non-traditional migrants with political power. Yet, it is this same cultural logic that current political leaders in Chan Kom exploit to legitimize their access to power. This access to power is deeply rooted in, and mobilized by, the migrants' experience within the tourist-market world. The Maya cultural complex, with its deep historic past and continual anthropological scrutiny during the twentieth century, has been consciously mobilized by a group of Maya, who monopolize political power for the practical purpose of placing Chan Kom on the tourist map.

Notes

1. After the long-term field work in Chan Kom in 1989–90, I have visited the community for short periods of time. From 1997 to 2003, I worked with Melinda Levin in an ethnographic documentary on the Chan Kom–Cancún relationship. This interview with the president of Chan Kom is ethnographic material from this documentary project.

2. *Primicia* is a general term under which Maya peasants identify different ceremonial performances to honor the supernatural forces that nurture and protect the *milpa*.

3. *Yuntziles* or *Yutziloob* is a general Maya term that refers to sacred entities that protect the village, milpas, and forest.

4. *X-Kukikan* derives from *Kukulkan*, a version of the central Mexican Quetzalcoatl, the sacred feathered serpent adopted by the Maya during the strong Nahuatl influence in the Postclassic period. "X" is the female morpheme in the Maya language, which means that Kukulkan is conceptualized as a female among contemporary Maya, probably because *serpiente* (the translation of "snake" into Spanish) is female.

Bibliography

Abrams, Elliot M.
  1994 *How the Maya Built Their World: Energetics and Ancient Architecture.* University of Texas Press, Austin.
Acosta, Jorge R.
  1945 La cuarta y quinta temporada de excavaciones en Tula, Hgo., 1943–1944. *Revista Mexicana de Estudios Antropológicos* 7:23–64.
Acosta, Jorge R., and Javier Romero
  1992 *Exploraciones en Monte Negro, Oaxaca 1937–38, 1938–39, 1939–40.* Antologías Serie Arqueología, comp. José Luis Ramírez R., coord. Lorena Mirambel. Instituto Nacional de Antropología e Historia, Mexico City.
Adams, Richard E. W.
  1986 Rio Azul. *National Geographic* 169:420–51.
  1999 *Rio Azul: An Ancient Maya City.* University of Oklahoma Press, Norman.
Adams, Richard E. W., and Woodruff D. Smith
  1981 Feudal Models for Classic Maya Civilization. In *Lowland Maya Settlement Patterns*, ed. Wendy A. Ashmore, pp. 335–49. University of New Mexico Press, Albuquerque.
Alexander, L. T., and J. G. Cady
  1962 *Genesis and Hardening of Laterites in Soils.* Technical Bulletin no. 1282. U.S. Department of Agriculture, Washington, DC.
Alexander, Rani
  1999 Mesoamerican Houselots and Archaeological Site Structure: Problems of Inference in Yaxcaba, Yucatán: Mexico, 1750–1847. In *The Archaeology of Household Activities*, ed. P. Allison, pp. 87–100. Routledge, London.
Ambrosino, James
  2003 The Function of a Maya Palace at Yaxuna: A Contextual Approach. In *Maya Palaces and Elite Residences: An Interdisciplinary Approach*, ed. J. J. Christie, pp. 253–73. University of Texas Press, Austin.
Ambrosino, James, and Kam Manahan
  1998 Investigations of the Perimeter of the North Acropolis. In *The Selz Foundation Yaxuná Project: Final Report of the 1996 Field Season*, ed. Justine M. Shaw and David A. Freidel, pp. 5–17. Department of Anthropology, Southern Methodist University, Dallas.
American Meteorological Society (AMS)
  1995 Hurricane Season Ends on Record Note. *AMS Newsletter* (December):7–8.
Anaya, Ana Luisa, Sergio Palacios Mayorga, and Eleazar González Velázquez
  1997 *El perifiton de la Reserva de El Eden, Quintana Roo, México: Un biomejorador potencial en la agricultural.* Paper presented at the El Edén Ecological Reserve Workshop, University of California, Riverside.
Andersen, Bente Juhl
  2001 *Pre-Hispanic Subsistence Strategies: A Comparison between El Eden and Other Selected Sites in the Maya Lowlands.* Master's thesis, Institute for Archaeology and Ethnology, University of Copenhagen.

Anderson, Benedict
  1983   *Imagined Communities: Reflections on the Origin and Spread of Nationalism.* Verso,
         London.
Anderson, David S., Anthony P. Andrews and Fernandon Robles Castellanos
  2004   *The Preclassic in Northwest Yucatan.* Paper presented at the 103d Annual Meet-
         ing of the American Anthropological Association, Atlanta.
Anderson, Eugene N.
  1993   Natural Resource Use in a Maya Village. In *The View from Yalahau: 1993 Ar-
         chaeological Investigations in Northern Quintana Roo, Mexico,* ed. S. L. Fedick and
         K. Taube, pp. 139–48. Latin American Studies Program Field Report Series,
         Vol. 2. University of California, Riverside.
Anderson, Patricia K.
  1998a Yula, Yucatán, Mexico: Terminal Classic Maya Ceramic Chronology for the
         Chichén Itzá Area. *Ancient Mesoamerica* 9:151–65.
  1998b *Yula, Yucatán, Mexico: Terminal Classic Maya Settlement and Political Organi-
         zation in the Chichen Itza Polity.* Ph.D. dissertation, Department of Anthro-
         pology, University of Chicago. University Microfilms, Ann Arbor, MI.
Andrews, Anthony P.
  1990a The Fall of Chichén Itzá: A Preliminary Hypothesis. *Latin American Antiquity*
         1(3):258–67.
  1990b The Role of Trading Ports in Maya Civilization. In *Vision and Revision in Maya
         Studies,* ed. Flora S. Clancy and Peter D. Harrison, pp. 159–68. University of
         New Mexico Press, Albuquerque.
Andrews, Anthony P., E. Wyllys Andrews V, and Fernando Robles Castellanos
  2003   The Northern Maya Collapse and Its Aftermath. *Ancient Mesoamerica* 14:151–
         56.
Andrews, Anthony P., Frank Asaro, Helen V. Michel, Fred H. Stross,
and Pura Cervera Rivero
  1989   The Obsidian Trade at Isla Cerritos, Yucatán, Mexico. *Journal of Field Archae-
         ology* 16:355–63.
Andrews, Anthony P., Tomás Gallareta Negrón, and Rafael Cobos Palma
  1989   Preliminary Report of the Cupul Survey Project. *Mexicon* 11:91–95.
Andrews, Anthony P., and Fernando Robles Castellanos
  1985   Chichén Itzá and Cobá: An Itzá-Maya Standoff in Early Postclassic Yucatán.
         In *The Lowland Maya Postclassic,* ed. Arlen Chase and Prudence Rice, pp. 62–
         72. University of Texas Press, Austin.
  2001   *Archaeological Survey of Northwest Yucatán.* Paper presented at the 66th Annual
         Meeting of the Society for American Archaeology, New Orleans.
Andrews, E. Wyllys, IV
  1970   *Balankanche, Throne of the Tiger Priest.* Middle American Research Institute
         Publication 32. Tulane University, New Orleans.
Andrews, E. Wyllis, IV, and Anthony P. Andrews
  1975   *A Preliminary Study of the Ruins of Xcaret, Quintana Roo, Mexico.* Middle Ameri-
         can Research Institute Publication 40. Tulane University, New Orleans.
Andrews, E. Wyllys, IV, and E. Wyllys Andrews V
  1980   *Excavations at Dzibilchaltún, Yucatán, Mexico.* Middle American Research In-
         stitute Publication 48. Tulane University, New Orleans.
Andrews, E. Wyllys, IV, and George E. Stuart
  1968   The Ruins of Ikil, Yucatán, Mexico. In *Archaeological Investigations on the Yuca-*

tán Peninsula, ed. Margaret Harrison and Robert Wauchope, pp. 69–80. Middle American Research Institute Publication 31. Tulane University, New Orleans.

Andrews, E. Wyllys, V

1979   Some Comments on Puuc Architecture of the Northern Yucatán Peninsula. In *The Puuc: New Perspectives*, ed. Lawrence Mills, pp. 1–17. Scholarly Studies in the Liberal Arts, Publication 1. Central College, Pella, IA.

1981   Dzibilchaltún. In *Archaeology*, ed. Jeremy A. Sabloff, pp. 313–44. Handbook of Middle American Indians, Supplement 1. University of Texas Press, Austin.

1986   Olmec Jades from Chacsinkin, Yucatan and Maya Ceramics from La Venta, Tabasco. In *Research and Reflections in Archaeology and History*, ed. E. W. Andrews V, pp. 11–49. Middle American Research Institute Publication 57. Tulane University, New Orleans.

1988   Ceramic Units from Komchen, Yucatán, Mexico. *Cerámica de Cultura Maya* 15:51–64.

1989   *The Ceramics of Komchen, Yucatán*. Unpublished manuscript on file, Middle American Research Institute, Tulane University, New Orleans.

1990   The Early Ceramic History of the Lowland Maya. In *Vision and Revision in Maya Studies*, ed. Florence Clancy and Peter Harrison, pp. 1–20. University of New Mexico Press, Albuquerque.

2003   *New Thoughts on Komchén and the Late Middle Preclassic*. Paper presented at the 2d Annual Tulane Maya Symposium, New Orleans.

Andrews, George F.

1985   Chenes-Puuc Architecture: Chronology and Cultural Interaction. In *Arquitectura y Arqueología: Metodologías en la Cronología de Yucatán. Etudes Mesoaméricaines, série II-8*. Centre d'études Méxicaines et Centraméricaines, Mexico City.

Ardren, Traci A.

1997   *The Politics of Place: Architecture and Cultural Change at the Xkanha Group, Yaxuná, Yucatán, Mexico*. Ph.D. dissertation, Department of Anthropology, Yale University.

2002   Conversations about the Production of Archaeological Knowledge and Community Museums at Chunchucmil and Kochol, Yucatán, Mexico. *World Archaeology* 34(2):379–400.

Ardren, Traci A., Bruce H. Dahlin, Travis Stanton, Scott Hutson, and Aline Magnoni

2000   *Archaeology and the Maya Landscape at Chunchucmil, Yucatán, Mexico*. Paper presented at the 99th Annual Meeting of the American Anthropological Association, San Francisco.

Ardren, Traci A., and Scott R. Hutson

2002   Ancient Maya Religious Practices at Chunchucmil and Yaxuná. *PARI Journal* 3(1):5–10.

Ardren, Traci A., and Dave Johnstone

1996   *A Middle Preclassic Ceremonial Structure from Yaxuná, Yucatán, Mexico*. Paper presented at the 61st Annual Meeting of the Society for American Archaeology, New Orleans.

Ardren, Traci A., Aline Magnoni, and David Hixson

2003   *The Nature of Urbanism at Ancient Chunchucmil*. Paper presented at the 2d Annual Tulane Maya Symposium, New Orleans.

Arnold, Dean
    2003   Review of Rattray's Teotihuacan: Ceramics, Chronology, and Cultural Trends.
            Latin American Antiquity 14:506–8.
Arnold, Jeanne, and Annabel Ford
    1980   A Statistical Examination of Settlement Patterns at Tikal, Guatemala. *American Antiquity* 45(4):713–26.
Arnold, Philip
    1990   The Organization of Refuse Disposal and Ceramic Production within Contemporary Mexican Houselots. *American Anthropologist* 92(1990):915–32.
Ashmore, Wendy A.
    1981   Some Issues of Method and Theory in Lowland Maya Settlement Archaeology.
            In *Lowland Maya Settlement Patterns*, ed. Wendy A. Ashmore, pp. 37–69. School
            of American Research, University of New Mexico Press, Albuquerque.
    1991   Site-planning Principles and Concepts of Directionality among the Ancient
            Maya. *American Antiquity* 2(3):199–226.
Ashmore, Wendy A., and Richard R. Wilk
    1988   Household and Community in the Mesoamerican Past. In *Household and Community in the Mesoamerican Past*, ed. Richard R. Wilk and Wendy A. Ashmore,
            pp. 1–28. University of New Mexico Press, Albuquerque.
Ashmore, Wendy A., and Gordon R. Willey
    1981   An Historical Introduction to the Study of Lowland Maya Settlement Patterns.
            In *Lowland Maya Settlement Patterns*, ed. Wendy A. Ashmore, pp. 3–18. School
            of American Research, University of New Mexico Press, Albuquerque.
Aveni, Anthony F., and Horst Hartung
    1986   Maya City Planning and the Calendar. *Transactions of the American Philosophical Society* 76(7).
Back, William
    1985   Hydrogeology of the Yucatán. In *Geology and Hydrogeology of the Yucatán and
            Quaternary Geology of Northeastern Yucatán Peninsula*, ed. W. C Ward, A. E. Weidie, and W. Back, pp. 99–124. New Orleans Geological Society, New Orleans.
Ball, Joseph W.
    1977   *The Archaeological Ceramics of Becán, Campeche, Mexico*. Middle American Research Institute Publication 43. Tulane University, New Orleans.
    1978   Archeological Pottery of the Yucatán-Campeche Coast. In *Studies in the Archaeology of Coastal Yucatán and Campeche, Mexico*. Middle American Research
            Institute Publication 46. Tulane University, New Orleans.
    1979   Ceramics, Culture History, and the Puuc Tradition: Some Alternative Possibilities. In *The Puuc: New Perspectives*, ed. Lawrence Mills, pp. 18–35. Scholarly
            Studies in the Liberal Arts, Publication 1. Central College, Pella, IA.
    1982   The Tancah Ceramic Situation: Cultural and Historical Insights from an Alternative Material Class. Appendix 1. In *On the Edge of the Sea: Mural Painting at
            Tancah-Tulum*, ed. A. Miller, pp. 105–13. Dumbarton Oaks Research Library
            and Collection, Washington, DC.
    1993   Pottery, Potters, Palaces, and Polities: Some Socioeconomic and Political Implications of Late Classic Maya Ceramic Industries. In *Lowland Maya Civilization in the Eighth Century A.D.*, ed. Jeremy A. Sabloff and John S. Henderson, pp.
            243–72. Dumbarton Oaks Research Library and Collection, Washington, DC.
    1994   Northern Maya Archaeology: Some Observations on an Emerging Paradigm.
            In *Hidden Among the Hills: Maya Archaeology of the Northwest Yucatán Peninsula*,

ed. Hanns J. Prem, 389–96. Acta Mesoamericana, Vol. 7. Verlag Von Fleming, Möckmühl, Germany.

Ball, Joseph W., and Jennifer T. Taschek
    1991   Late Classic Lowland Maya Political Organization and Central-Place Analysis: New Insights from the Upper Belize Valley. *Ancient Mesoamerica* 2:149–65.

Barrera Marín, Alfredo
    1980   Sobre la unidad habitación tradicional campesina y el manejo de los recursos bióticos en el area Maya Yucatánense. Árboles y arbustos de los huertos familiares. *Biótica* 2(2):47–61.

Barrera Rubio, Alfredo
    1985   Settlement Patterns in Uxmal Area, Yucatán, Mexico. *Indiana* 10:227–35.

Barrera Rubio, Alfredo, and José Huchim Herrera
    1989   Exploraciones recientes en Uxmal (1986–1987). In *Memorias del Segundo Coloquio Internacional de Mayistas, 1*, ed. Alain Breton, pp. 265–86. Universidad Nacional Autónoma de México, Instituto de Investigaciones Filologicas, Mexico City.

Beach, Timothy
    1998   Soil Constraints on Northwest Yucatán, Mexico: Pedoarchaeology and Maya Subsistence at Chunchucmil. *Geoarchaeology* 13(8):759–91.

Beach, Timothy, and Nicholas P. Dunning
    1995   Ancient Maya Terracing and Modern Conservation in the Peten Rain Forest of Guatemala. *Journal of Soil and Water Conservation* 50:138–45.

Becker, Marshall
    1991   Plaza Plans at Tikal, Guatemala and at Other Lowland Maya Sites: Evidence for Patterns of Culture Change. *Cuadernos de Arquitectura Mesoamericana* 14:11–26.

Beekman, Christopher S.
    1996   Political Boundaries and Political Structure: The Limits of the Teuchitlan Tradition. *Ancient Mesoamerica* 7:135–47.

Bell, Julie
    1998   *A Developing Model for Determining Cenote and Associated Site Settlement Patterns in the Yalahau Region, Quintana Roo, Mexico.* Master's thesis, University of California, Riverside.

Benavides Castillo, Antonio
    1975   Cobá: Sus sacbeob y dzib mul. *INAH Boletin* II(15):55–58.
    1981   Cobá y Tulum: Adaptacion al medio ambiente y control del medio social. *Estudios de Cultura Maya* 13:205–22.

Benavides Castillo, Antonio, and Linda Manzanilla
    1987   Introducción: estudio de centros urbanos. In *Coba, Quintana Roo: análisis de dos unidades habitacionales Mayas*, ed. Linda Manzanilla, pp. 11–23. Instituto Nacional de Antropología e Historia, Mexico City.

Bey, George J., III, Tara M. Bond, William M. Ringle, Craig A. Hanson, Charles W. Houck, and Carlos Peraza Lope
    1998   The Ceramic Chronology of Ek Balam, Yucatán, Mexico. *Ancient Mesoamerica* 9:101–20.

Bey, George J., III and Rossana May Ciau
    2005   *Los orígenes preclásicos de un Centro Puuc en el Distrito de Bolonchén: Las evidencias de Kiuic.* Paper presented at the 2d Congreso Internacional de Cultura Maya, Mérida, Yucatán, Mexico.

Bey, George J., III, Craig A. Hanson, and William M. Ringle
    1997   Classic to Postclassic at Ek Balam, Yucatán: Architectural and Ceramic Evi-
            dence for Defining the Transition. *Latin American Antiquity* 8(3):237–54.
Bey, George J., III, and William M. Ringle
    1989   *The Myth of the Center: Political Integration at Ek Balam, Yucatán, Mexico*. Paper
            presented at the 54th Annual Meeting of the Society for American Archae-
            ology, Atlanta.
Binford, Lewis
    1982   The Archaeology of Place. *Journal of Anthropological Archaeology* 1:5–31.
Blackmore, Chelsea, and Traci A. Ardren
    2001   Excavations at the Pich Group. In *Pakbeh Regional Economy Program: Report of
            the 2001 Field Season*, ed. Bruce Dahlin and Daniel Mazeau, pp. 58–68. Depart-
            ment of Sociology and Anthropology, Howard University, Washington, DC.
Bond-Freeman, Tara M., George J. Bey III, Charles W. Houck, and William M. Ringle
    1998   *Ceramic Evidence from the Middle Preclassic Period at Ek Balam and Xuilub, Yuca-
            tán, Mexico*. Paper presented at the 97th Annual Meeting of the American
            Anthropological Association, Philadelphia.
Bond-Freeman, Tara M., George J. Bey III, and William M. Ringle
    2003   *A Paucity and a Plethora: The Early Classic Period Bichrome Ceramic Tradition at
            Ek Balam*. Unpublished manuscript, Southern Methodist University, Dallas.
Bond-Freeman, Tara M., George J. Bey III, William M. Ringle, and J. Gregory Smith
    1999   *The Ceramics of Ichmul de Morley, Yucatán, Mexico*. Paper presented at the 64th
            Annual Meeting of the Society for American Archaeology, Chicago.
Boucher, Sylviane
    1991   *El Preclásico Medio en el Puuc: Nuevos datos*. Articulo leído en la III Mesa Re-
            donda de la Sociedad Española de Estudios Mayas. Gerona, Spain.
    1996   *Preliminary Ceramic Report for the 1996 Field Season at El Naranjal, Quintana Roo,
            Mexico*. Ceramoteca Regional, Instituto Nacional de Antropología e Historia,
            Mérida, Yucatán, Mexico.
Brainerd, George W.
    1958   *The Archaeological Ceramics of Yucatán*. Anthropological Records Vol. 19. Uni-
            versity of California Press, Berkeley.
Braswell, Geoffrey E.
    1997   El intercambio prehispánico en Yucatán, México. In *X Simpósio de Investigacio-
            nes Arqueológicos en Guatemala*, ed. Juan Pedro Laport and Héctor L. Escobedo,
            vol. 2, pp. 545–56. Museo Nacional de Arqueología y Etnología, Guatemala.
    1998   *Trade and Procurement: Obsidian and the Maya of the Northern Lowlands*. Paper
            presented at the 97th Annual Meeting of the American Anthropological As-
            sociation, Philadelphia.
    2003   Introduction: Reinterpreting Early Classic Interaction. In *The Maya and Teoti-
            huacan, Reinterpreting Early Classic Interaction*, ed. Geoffrey E. Braswell, pp. 1–
            43. University of Texas Press, Austin.
Brinton, Daniel G.
    1882   *The Maya Chronicles*. Brinton's Library of Aboriginal American Literature 1.
            Philadelphia.
Britton, Stephen G.
    1982   The Political Economy of Tourism in the Third World. *Annals of Tourism Re-
            search* 9:331–58.

Bronson, Bennet
    1966   Roots and the Subsistence of the Ancient Maya. *Southwestern Journal of Anthropology* 22:251–79.
Brown, Denise Fay
    1999   Mayas and Tourists in the Maya World. *Human Organization* 58(3):295–304.
Bullard, William R., Jr.
    1952   *Residential Property Walls at Mayapan.* Current Reports No. 3. Department of Archaeology, Carnegie Institution of Washington, Washington, DC.
    1962   Settlement Pattern and Social Structure in the Southern Maya Lowlands during the Classic Period. In *Proceedings of XXXV Congreso Internacional de Americanistas*, pp. 279–87. Mexico City.
Burgos, Rafael, José Estrada, and Juan García
    2003   Una aproximación al patrón de asentamiento del sitio de Izamal, Yucatán. *Los Investigadores de la Cultura Maya* 11, 2:313–24. Universidad Autónoma de Campeche, Campeche, Mexico.
Caballero, Juan J.
    1988   *The Maya Homegardens of the Yucatán Peninsula: A Regional Comparative Study 1, 2.* Paper presented at the International Congress of Ethnobiology, Belem, Brazil.
Caldwell, Joseph R.
    1959   The New American Archaeology. *Science* 129(3345):303–7.
    1964   Interaction Spheres in Prehistory. *Illinois State Museum Scientific Papers* 12(6):133–56.
Canché, Elena
    1992   *La secuencia cerámica de Xelha, Quintana Roo.* Master's thesis, Universidad Autónoma de Yucatán, Mérida, Yucatán, Mexico.
Carmona Macias, Martha
    1993   Catalogue of Objects. In *Teotihuacan: Art from the City of the Gods*, ed. K. Berrin and E. Pasztrory, p. 271. Thames and Hudson, NY.
Carrasco, David
    2000   *Quetzalcoatl and the Irony of Empire: Myths and Prophecies in the Aztec Tradition.* University of Colorado Press, Boulder.
Castañeda, Quetzil E.
    1996   *In the Museum of Maya Culture: Touring Chichén Itzá.* University of Minnesota Press, Minneapolis.
Ceballos Gallareta, Teresa
    2004   *Informe preliminar de la cerámica del sitio arqueológico de Ek' Balám, Yucatán.* Archive de la Seccion de Arqueologia del Centro INAH Yucatán, Mérida, Yucatán, Mexico.
Chang, Jen-Hu
    1968   The Agricultural Potential of the Humid Tropics. *Geographical Review* 58:333–61.
Charnay, Désiré
    1887   *The Ancient Cities of the New World: Being Voyages and Explorations in Mexico and Central America from 1857–1882.* Trans. J. Gonine and Helen S. Conant. Chapman, London 1987; Harper Brothers, NY, 1987.
Chase, Arlen F., and Diane Z. Chase
    1996   More than Kin and King: Centralized Political Organization among the Late Classic Maya. *Current Anthropology* 37:803–10.

Chase, Diane Z., and Arlen F. Chase
   1982   Yucatec Influence in Terminal Classic Northern Belize. *American Antiquity* 47:
          596–614.
   1992   An Archaeological Assessment of Mesoamerican Elites. In *Mesoamerican Elites:
          An Archaeological Assessment*, ed. Diane Z. Chase and Arlen F. Chase, pp. 303–
          17. University of Oklahoma Press, Norman.
Chippindale, Christopher
   2000   Capta and Data: On the True Nature of Archaeological Information. *American
          Antiquity* 65:605–12.
Clancy, Flora S.
   1985   Maya Iconography. In *Maya: Treasures of an Ancient Civilization*, ed. C. Gallen-
          kamp and E. E. Johnson, pp. 58–70. Harry N. Abrams, NY.
Clendinnen, Inga
   1987   *Ambivalent Conquests: Maya and Spaniard in Yucatán, 1517–1570*. Cambridge
          Latin American Studies. Cambridge University Press, Cambridge.
Cobean, Robert H.
   1990   *La cerámica de Tula, Hidalgo*. Colección Científica número 215, Serie Arqueo-
          logía. Instituto Nacional de Antropología e Historia, Mexico City.
Cobean, Robert H., and Alba Guadalupe Mastache
   1987   Cerámica importada en Tula, Hidalgo: Un informe preliminar. *Arqueología*
          1:89–132.
Cobos, Rafael
   1997   Katún y ahau: fechando el fin de Chichén Itzá. In *Identidades Sociales en Yuca-
          tán*, ed. María Cecilia Lara C., pp. 19–40. Facultad de Ciencias Antropológicas,
          Universidad Autónoma de Yucatán, Mérida, Yucatán, Mexico.
   1998   *Chichén Itzá: Análisis de una comunidad del Período Clásico Terminal*. Los Inves-
          tigadores de la Cultura Maya 6, 2:316–31. Universidad Autónoma de Cam-
          peche, Campeche, Mexico.
   2000   *Multepal or Centralized Kingship?: New Evidence on Governmental Organization
          at Chichén Itzá*. Paper presented at the colloquium *Rethinking Tula, Tollan, and
          Chichén Itzá*. Dumbarton Oaks Research Library and Collection, Washington,
          DC.
   2001   El centro de Yucatán: De area periférica a la integración de la comunidad ur-
          bana en Chichén Itzá. In *Reconstruyendo la Ciudad Maya: El urbanismo en las
          sociedades antiguas*, ed. Andrés Ciudad Ruiz, Ma. Josefa Iglesias Ponce de León,
          and María del Carmen Martínez Martínez, pp. 253–76. Publicaciones de la
          Sociedad Española de Estudios Mayas, Número 6. Madrid.
   2003a Ancient Community Form and Social Complexity at Chichén Itzá. In *Urban-
          ism in Mesoamerica*, ed. William T. Sanders, Alba Guadalupe Mastache, and
          Robert H. Cobean, 1:451–72. Instituto Nacional de Antropología e Historia,
          Mexico City; The Pennsylvania State University, University Park.
   2003b *The Settlement Patterns of Chichén Itzá, Yucatán*. Ph.D. dissertation, Department
          of Anthropology, Tulane University, New Orleans.
   2004   Chichén Itzá, Settlement and Hegemony during the Terminal Classic Period.
          In *The Terminal Classic in the Maya Lowlands, Collapse, Transition, and Transfor-
          mation*, ed. Arthur A. Demarest, Prudence M. Rice, and Don S. Rice, pp. 517–
          44. University Press of Colorado, Boulder.
   2005   *Los moluscos arqueológicos de Yucatán durante el Preclásico*. Paper presented at
          the 2d Congreso Internacional de Cultura Maya, Mérida, Yucatán, Mexico.

Cobos, Rafael, and Terance L. Winemiller
    2001   The Late and Terminal Classic-Period Causeway Systems of Chichén Itzá, Yucatán, Mexico. *Ancient Mesoamerica* 12:283–91.
Coe, Michael D.
    1961   Social Typology and Tropical Forest Civilizations. *Comparative Studies in Society and History* 4(1):65–85.
    1999   *Breaking the Maya Code*. Rev. ed. Thames and Hudson, London.
Coggins, Clemency, and John M. Ladd
    1992   Wooden Artifacts. In *Artifacts from the Cenote of Sacrifice, Chichén Itzá, Yucatán*, ed. Clemency Coggins, pp. 235–344. Memoirs of the Peabody Museum, 10:3. Harvard University, Cambridge, MA.
Coke, James, E. C. Perry, and A. Long
    1991   Charcoal from a Probable Fire Pit on the Yucatán Peninsula, Mexico: Another Point on the Glacio-Eustatic Sea Level Curve. *Nature* 353:25.
Con, Maria José, and Alejandro Martinez Muriel
    2001   Cobá: Amid Roads and Lakes. *Arqueología Mexicana* 9(54):92–93.
Covich, A. P., and N. H. Nickerson
    1966   Studies of Cultivated Plants in Choco Dwelling Clearings, Darien, Panama. *Economic Botany* 20:285–301.
Cowgill, George L.
    2003   Teotihuacan and Early Classic Interaction: A Perspective from Outside the Maya Region. In *The Maya and Teotihuacan, Reinterpreting Early Classic Interaction*, ed. Geoffrey E. Braswell, pp. 315–35. University of Texas Press, Austin.
Cózatl Manzano, Roberto
    1999   *Sistemática de los moluscos dulceacuícolaslas de la Reserva Ecologica El Edén, Quintana Roo, México*. Bachelor's thesis, Universidad Nacional Autónoma de México, Mexico City.
Culbert, T. Patrick
    1991   Polities in the Northeast Peten, Guatemala. In *Classic Maya Political History: Hieroglyphic and Archaeological Evidence*, ed. T. Patrick Culbert, pp. 128–46. Cambridge University Press, Cambridge.
Culbert, T. Patrick, Laura J. Levi, and Luis Cruz
    1990   Lowland Maya Wetland Agriculture: The Rio Azul Agronomy Program. In *Vision and Revision in Maya Studies*, ed. Flora Clancy and Peter Harrison, pp. 115–24. University of New Mexico Press, Albuquerque.
Dahlin, Bruce H.
    2000   The Barricade and Abandonment of Chunchucmil: Implications for Northern Maya Warfare. *Latin American Antiquity* 11:283–98.
    2001   Operation 15: A Possible Marketplace. In *Pakbeh Regional Economy Program: Report of the 2001 Season*, ed. Bruce H. Dahlin and Daniel Mazeau, pp. 20–25. Department of Sociology and Anthropology, Howard University, Washington, DC.
Dahlin, Bruce H., Anthony Andrews, Tim Beach, Clara Bezanilla, Pat Farrell, Sheryl Luzzader-Beach, and Valerie McCormick
    1998   Punta Canbalam in Context: A Peripatetic Coastal Site in Northwest Campeche, Mexico. *Ancient Mesoamerica* 9:1–15.
Dahlin, Bruce H., and Traci A. Ardren
    2002   Modes of Exchange and Regional Patterns: Chunchucmil, Yucatán, Mexico.

In *Ancient Maya Political Economies*, ed. Marilyn Masson and David Freidel, pp. 249–84. Altamira, Walnut Creek, CA.

Dahlin, Bruce H., Traci A. Ardren, Aline Magnoni, and Scott Hutson
1999   *The Peculiar Cultural Landscape of Chunchucmil*. Poster presented at the 64th Annual Meeting of the Society for American Archaeology, Chicago.

de la Garza, Mercedes (ed.)
1983   *Relaciones histórico-geográficas de la gobernación de Yucatán*. Instituto de Investigaciones Filológicas, Centro de Estudios Mayas, Universidad Nacional Autónoma de México, Mexico City.

de Landa, Fray Diego
1959   *Relación de las cosas de Yucatán*. Editorial Porrúa, Mexico City.

de Montmollin, Olivier
1989   *The Archaeology of Political Structure: Settlement Analysis in a Classic Maya Polity*. Cambridge University Press, Cambridge.
1995   *Settlement and Politics in Three Late Classic Maya Polities*. Prehistory Press, Madison, WI.

de Pierrebourg, Fabienne
1999   *Une approche ethnoarchéologique au Yucatán (Mexique)*. Paris Monographs in American Archaeology 3. BAR S764. Hadrian Books, London.

Demarest, Arthur A.
1992   Ideology in Ancient Maya Cultural Evolution: The Dynamics of Galactic Polities. In *Ideology and Pre-Columbian Civilizations*, ed. Arthur A. Demarest and Geoffrey W. Conrad, pp. 135–57. School of American Research Press, Santa Fe, NM.
1993   The Violent Saga of a Maya Kingdom. *National Geographic* 183(2):94–111.
2004   *Ancient Maya: The Rise and Fall of a Rainforest Civilization*. Cambridge University Press, Cambridge.

Demarest, Arthur A., and Antonia E. Foias
1993   Mesoamerican Horizons and the Cultural Transformations of Maya Civilization. In *Latin American Horizons*, ed. Don S. Rice, pp. 146–91. Dumbarton Oaks Research Library and Collection, Washington, DC.

Denevan, W. M., and K. Schwerin
1978   Adaptive Strategies in Karinya Subsistence, Venezuelan Llanos. *Antropología* (50):3–91.

Díaz del Castillo, Bernal
1956   *Bernal Díaz Chronicles: The True Story of the Conquest of Mexico*. Doubleday, Garden City, NJ.

Diehl, Richard A.
1981   Tula. In *Archaeology*, ed. Jeremy A. Sabloff, pp. 277–95. Supplement to the Handbook of Middle American Indians, vol. 1, Victoria R. Bricker, general ed. University of Texas Press, Austin.
1983   *Tula: The Toltec Capital of Ancient Mexico*. Thames and Hudson, London.
1993   The Toltec Horizon in Mesoamerica: New Perspectives on an Old Issue. In *Latin American Horizons*, ed. Don S. Rice, pp. 263–94. Dumbarton Oaks Research Library and Collection, Washington, DC.

Dongoske, Kurt E., Mark Aldenderfer, and Karen Doehner (eds.)
2000   *Working Together: Native Americans & Archaeologists*. Society for American Archaeology, Washington, DC.

Downum, Christian E., and Laurie J. Price
    1999   Applied Archaeology. *Human Organization* 58(3):226–39.
Dunham, Peter S.
    1990   *Coming Apart at the Seams: The Classic Development and Demise of Maya Civili-zation: A Segmentary View from Xnaheb, Belize.* Ph.D. dissertation, Department of Anthropology, State University of New York at Albany. University Micro-films, Ann Arbor, MI.
Dunning, Nicholas P.
    1992   *Lords of the Hills: Ancient Maya Settlement in the Puuc Region, Yucatán, Mexico.* Prehistory Press, Madison, WI.
Dunning, Nicholas P., and Timothy Beach
    2000   Stability and Instability in Prehistoric Maya Landscapes. In *Imperfect Balance: Landscape Transformation in the Precolumbian Americas,* ed. D. L. Lentz, pp. 179–202. Columbia University Press, NY.
Dunning, Nicholas P., and Jeff Karl Kowalski
    1994   Lords of the Hills: Classic Maya Settlement Patterns and Political Iconography in the Puuc Region, Mexico. *Ancient Mesoamerica* 5:63–95.
Dutton, Bertha
    1952   *The Toltecs and Their Influence on the Culture of Chichén Itzá.* Ph.D. dissertation, Department of Anthropology, Columbia University. University Microfilms, Ann Arbor, MI.
Elmendorf, Mary
    1972   *Mayan Women and Change.* Cuaderno No. 81. CIDOC, Cuernavaca, Mexico.
    1976   *Nine Maya Women: A Village Faces Change.* Schenkman Pub. Co., NY.
Elmendorf, Mary, and Devorah Merrill
    1977   *Socioeconomic Impact of Development in Chan Kom, 1971–1976: Rural Women Participate in Change.* Unpublished manuscript prepared for the World Bank Population and Human Resources Division, Development Economics Depart-ment.
    1978   *Socioeconomic Impact of Development in Chan Kom, Yucatán, 1971–1976: A Pre-liminary Study.* Paper presented at the 1987 Meeting of the Society for Applied Anthropology, Mérida, Yucatán, Mexico.
Evans, Susan
    1993   Aztec Household Organization and Village Administration. In *Prehispanic Do-mestic Units in Western Mesoamerica,* ed. Robert S. Santley and Kenneth Hirth, pp. 173–90. CRC Press, Boca Raton, FL.
Farriss, Nancy
    1984   *Maya Society Under Colonial Rule: The Collective Enterprise of Survival.* Princeton University Press, Princeton, NJ.
Fedick, Scott L.
    1996a   Conclusion: Landscape Approaches to the Study of Ancient Maya Agriculture and Resource Use. In *The Managed Mosaic: Ancient Maya Agriculture and Re-source Use,* ed. Scott L. Fedick, pp. 355–47. University of Utah Press, Salt Lake City.
    1996b   An Interpretative Kaleidoscope: Alternative Perspectives on Ancient Agricul-tural Landscapes of the Maya Lowlands. In *The Managed Mosaic: Ancient Maya Agriculture and Resource Use,* ed. Scott L. Fedick, pp. 107–31. University of Utah Press, Salt Lake City.

1996c New Perspectives on Ancient Maya Agriculture and Resource Use. In *The Man-aged Mosaic: Ancient Maya Agriculture and Resource Use*, ed. Scott L. Fedick, pp. 1–16. University of Utah Press, Salt Lake City.

1998a *El Proyecto Yalahau: Informe técnico final sobre las investigaciones arqueológicas de 1996–1997 en el norte de Quintana Roo, México*. Report prepared for the Consejo de Arqueología of the Instituto Nacional de Antropología e Historia, Mexico City.

1998b Ancient Maya Use of Wetlands in Northern Quintana Roo, Mexico. In *Hidden Dimensions: The Cultural Significance Of Wetland Archaeology*, ed. K. Bernick, pp. 107–29. University of British Columbia Press, Vancouver, British Columbia, Canada.

2003 Archaeological Evidence for Ancient and Historic Resource Use Associated with the El Edén Wetland, Northern Quintana Roo, Mexico. In *The Lowland Maya Area: Three Millennia at the Human-Wildland Interface*, ed. Arturo Gómez-Pompa, Michael F. Allen, Scott L. Fedick, and Juan J. Jiménez-Osornio, pp. 339–60. Haworth Press, Binghamton, NY.

Fedick, Scott L., and Kevin Hovey

1995 Ancient Maya Settlement and Use of Wetlands at Naranjal and the Surrounding Yalahau Region. In *The View from Yalahau: 1993 Archaeological Investigations in Northern Quintana Roo, Mexico*, ed. Scott L. Fedick and Karl A. Taube. Field Report Series, No. 2. Latin American Studies Program, University of California, Riverside.

Fedick, Scott L., Bethany A. Morrison, Bente J. Andersen, Sylviane Boucher, Jorge Ceja Acosta, and Jennifer P. Mathews

2000 Wetland Manipulation in the Yalahau Region of the Northern Maya Lowlands. *Journal of Field Archaeology* 27:131–52.

Fedick, Scott L., and Karl A. Taube (eds.)

1995 *The View from Yalahau: 1993 Archaeological Investigations in Northern Quintana Roo, Mexico*. Latin American Studies Program, Field Report Series 2. University of California, Riverside.

Fernández Souza, Lilia, Rafael Cobos, and Marisa Vazquez de Agrdos

2002 Arquitectura en Sihó, Yucatán: la Estructura 5D16. In *Los Investigadores de la Cultura Maya*, No. 11, 2:355–61. Universidad Autónoma de Campeche, Campeche, Mexico.

Fish, Suzanne K., and Stephen A. Kowalewski (eds.)

1990 *The Archaeology of Regions: A Case for Full Coverage Survey*. Smithsonian Institution Press, Washington, DC.

Fletcher, Laraine

1983 Cobá and Mayapán: A Comparison of Solares, Houselot Variation, Sociopolitical Organization and Land Tenure. In *Cobá: A Classic Maya Metropolis*, ed. William J. Folan, Ellen R. Kintz, and Laraine A. Fletcher, pp. 121–31. Academic Press, NY.

Flores Peña, Sergio

1993 Los servicios públicos en las culturas prehispánicas: Un nuevo punto en la agenda de investigaciones. *Cuadernos de Arquitectura Mesoamericana* 23:59–63.

Folan, William J.

1992 Calakmul, Campeche: A Centralized Urban Administrative Center in Northern Petén. *World Archaeology* 24(1):158–68.

Folan, William J., Ellen R. Kintz, and Laraine A. Fletcher
  1983   *Coba: A Classic Maya Metropolis.* Academic Press, NY.
Ford, Anabel
  1990   Maya Settlement in the Belize River Area: Variations in Residence Patterns of
         the Central Maya Lowlands. In *Precolumbian Population History in the Maya
         Lowlands*, ed. T. Patrick Culbert and Don S. Rice, pp. 167–81. University of New
         Mexico Press, Albuquerque.
  1991   Evidence of Economic Variation of Ancient Maya Residential Settlement in
         the Upper Belize River Area. *Ancient Mesoamerica* 2:35–45.
  1999   *El Pilar Landscape: The Past Informs the Future.* Paper presented at the 64th An-
         nual Meeting of the Society for American Archaeology, Chicago.
Fox, John W.
  1987   *Maya Postclassic State Formation: Segmentary Lineage Migration in Advancing
         Frontiers.* Cambridge University Press, Cambridge.
Fox, John W., Garret W. Cook, Arlen F. Chase, and Diane Z. Chase
  1996   The Maya State: Centralized or Segmentary? *Current Anthropology* 37:795–830.
Freidel, David A.
  1979   Culture Areas and Interaction Spheres: Contrasting Approaches to the Emer-
         gence of Civilization in the Maya Lowlands. *American Antiquity* 44(1):36–54.
  1986   *The Yaxuná Archaeological Survey: A Study of the Battlefield in the Wars Between
         Chichén Itzá and Cobá and an Investigation of Large Scale State Formation in Low-
         land Maya Civilization.* Proposal Submitted to the National Endowment for
         the Humanities, Washington, DC.
  1992   Children of the First Father's Skull: Terminal Classic Warfare in the Northern
         Maya Lowlands and the Transformation of Kingship and Elite Hierarchies.
         In *Mesoamerican Elites: An Archaeological Assessment*, ed. Diane Z. Chase and
         Arlen F. Chase, pp. 99–117. University of Oklahoma Press, Norman.
Freidel, David A., and Jeremy A. Sabloff
  1984   *Cozumel: Late Maya Settlement Patterns.* Academic Press, NY.
Freidel, David A., and Charles Suhler
  1998   Visiones Serpentinas y Laberintos Mayas. *Arqueología Méxicana* 6(34):28–37.
  1999   The Path of Life: Toward a Functional Analysis of Ancient Maya Architecture.
         In *Mesoamerican Architecture as a Cultural Symbol*, ed. Jeff K. Kowalski, pp. 250–
         73. Oxford University Press, Oxford.
Freidel, David A., Charles Suhler, and Rafael Cobos (eds.)
  1992   *The Selz Foundation Yaxuná Project. Final Report of the 1992 Field Season.* Depart-
         ment of Anthropology, Southern Methodist University, Dallas.
Freidel, David A., Charles Suhler, and Ruth Krochock
  1990   *Yaxuná Archaeological Survey: A Report of the 1989 Field Season and Final Report of
         Phase I.* Department of Anthropology, Southern Methodist University, Dallas.
Fry, Robert E.
  1969   *Ceramics and Settlement in the Periphery of Tikal, Guatemala.* Ph.D. dissertation,
         Department of Anthropology, University of Arizona.
  1979   The Economics of Pottery at Tikal, Guatemala: Models of Exchange for Serving
         Vessels. *American Antiquity* 44:494–512.
  1980   Models of Exchange for Major Shape Classes of Lowland Maya Pottery. In
         Models and Methods in Regional Exchange, ed. R. E. Fry. SAA Papers No. 1. So-
         ciety for American Archaeology, Washington, DC.

Gallareta Ceballos, Teresa, and Socorro Jimenez Alvarez
n.d.   Las esferas cerámicas del horizonte Cochuah del Clásico Temprano (c. 250–
        600 DC.) en el norte de la Peninsula de Yucatán. In *La Produccion Alfarera en
        el México Antiguo*, coords. Norberto Gonzalez Crespo and Angel Garcia Cook.
        Instituto Nacional de Antropología e Historia, Mexico City.
Gallareta Negrón, Tomás
1991   *Ichmul de Morley: Informe del reconocimiento realizado en Ichmul de Morley, Yuca-
        tán, México.* Manuscript on file at the Instituto Nacional de Antropología e
        Historia, Mérida, Yucatán, Mexico.
Gallareta Negrón, Tomás, George J. Bey III, and William M. Ringle
2002   *Investigaciones arqueológicas en las ruinas de Kiuic y la zona Labná-Kiuic, Distrito
        de Bolonchén, Yucatán, México, temporada 2001.* Report prepared for the Con-
        sejo de Arqueología del Instituto Nacional de Antropología e Historia, Mexico
        City.
2003   *Investigaciones arqueológicas en las ruinas de Kiuic y la zona Labná-Kiuic, Distrito
        de Bolonchén, Yucatán, México, temporada 2002.* Report prepared for the Con-
        sejo de Arqueología del Instituto Nacional de Antropología e Historia, Mexico
        City.
2004   *Investigaciones arqueológicas en las ruinas de Kiuic y la zona Labná-Kiuic, Distrito
        de Bolonchén, Yucatán, México, temporada 2003.* Report prepared for the Con-
        sejo de Arqueología del Instituto Nacional de Antropología e Historia, Mexico
        City.
Gallareta Negrón, Tomás, and William M. Ringle
2004   *The Earliest Occupation of the Puuc Region, Yucatan, Mexico: New Perspectives from
        Xocnaceh and Paso del Macho.* Paper presented at the 103d Annual Meeting of
        the American Anthropological Association, Atlanta.
Gallareta Negrón, Tomás, and Karl A. Taube
2005   Late Postclassic Occupation in the Ruinas de San Angel Region, Quintana Roo,
        Mexico. In *Quintana Roo Archaeology*, ed. Justine M. Shaw and Jennifer P.
        Mathews. University of Arizona Press, Tucson.
Gallareta Negrón, Tomás, Lourdes Toscano Hernández, Carlos Pérez Álvarez,
and Carlos Peraza Lope
1999   Proyecto Labná, Yucatán, México. In *Land of the Turkey and the Deer: Recent
        Archaeological Research in Yucatán*, ed. Ruth Gubler, pp. 85–96. Labyrinthos,
        Lancaster, CA.
Gann, Thomas W. F.
1925   *Mystery Cities: Exploration and Adventury in Lubaantun.* Scribner, London.
Garza Tarazona de González, Silvia, and Edward B. Kurjack
1980   *Atlas arqueológico del estado de Yucatán.* Instituto Nacional de Antropología e
        Historia, Mexico City.
Geertz, Clifford
1980   *Negara: The Theatre State in Nineteenth-Century Bali.* Princeton University Press,
        Princeton, NJ.
Gillespie, Susan D.
1989   *The Aztec Kings: The Construction of Rulership in Mexican History.* University of
        Arizona Press, Tucson.
Gleissman, S. R., B. L. Turner II, F. J. Rosado May, and M. F. Amador
1983   Raised-Field Agriculture in the Maya Lowlands of Southeastern Mexico. In

Drained Field Agriculture in Central and South America, ed. J. P. Darch, pp. 91–110. BAR International Series, No. 250. British Archaeological Reports, Oxford.

Glover, Jeffrey B., and Fabio Esteban Amador

2002 Reconocimiento Arqueológico en la Región Yalahau. In *Proyecto Regional de Ecología Humana: Informe Técnico de las Investigaciones Arqueológicas 2001 en T'isil y Reconocimiento en la Región de Yalahau, en el Norte de Quintana Roo, México*. Report prepared for the Consejo de Arqueología del Instituto Nacional de Antropología e Historia, Mexico City.

2004 Reconocimiento Arqueológico en la Región Yalahau. In *Proyecto Regional de Ecología Humana: Informe Técnico de las Investigaciones Arqueológicas 2003 en T'isil y Reconocimiento en la Región de Yalahau, en el Norte de Quintana Roo, México*. Report prepared for the Consejo de Arqueología del Instituto Nacional de Antropología e Historia, Mexico City.

2005 Recent Research in the Yalahau Region of Quintana Roo: Methodological Concerns and Preliminary Results of a Regional Survey. In *Quintana Roo Archaeology*, ed. Justine M. Shaw and Jennifer P. Mathews. University of Arizona Press, Tucson.

Goldkind, Victor

1965 Social Stratification in the Peasant Community: Redfield's Chan Kom Reinterpreted. *American Anthropologist* 67:863–84.

1966 Class Conflict and Cacique in Chan Kom. *Southwestern Journal of Anthropology*. 22:325–45.

Gómez-Pompa, Arturo, José Salvador Flores, and Mario Aliphat Fernández

1990 The Sacred Cacao Groves of the Maya. *Latin American Antiquity* 1:247–57.

Goñi Motilla, Guillermo A.

1993 *Solares prehispánicos en la Peninsula de Yucatán*. Bachelor's thesis, Escuela Nacional de Antropología e Historia, Mexico City.

1998 *Xamanhá: Un Sitio Arqueológico de la Costa Central de Quintana Roo*. Serie Arqueología. Instituto Nacional de Antropología e Historia, Mexico City.

Gorenstein, Shirley

1985 *Acambaro: Frontier Settlement on the Tarascan-Aztec Border*. Publications in Anthropology No. 32. Vanderbilt University, Nashville.

Graña-Behrens, Daniel

2002 *Die Maya-Inschriften aus Nordwestyukatan, Mexiko*. Inaugural Ph.D. dissertation, Rheinischen Friedrich-Wilhems-Universität zu Bonn, Bonn, Germany.

Greene Robertson, Merle

1993 *Merle Greene Robertson's Rubbings of Maya Sculpture*. CD-ROM set. Vol. V. Pre-Columbian Art Research Institute, San Francisco.

Greenwood, David

1989 Culture by the Pound: An Anthropological Perspective on Tourism as Cultural Commoditization. In *Hosts and Guests: The Anthropology of Tourism*, ed. Valene L. Smith, 2d ed., pp. 171–85. University of Pennsylvania Press, Philadelphia.

Griffith, Cameron, and Chip Colwell

2000 *Beyond the Ivy-Covered Walls: An Examination of Applied Archaeology in Belize*. Paper presented at the 99th Annual Meeting of the American Anthropological Association, San Francisco.

Grube, Nikolai

1994 Hieroglyphic Sources for the History of Northwest Yucatán. In *Hidden Among the Hills: Maya Archaeology of the Northwest Yucatán Peninsula*, ed. H. J. Prem. Acta Mesoamericana 7. Verlag Von Flemming, Möckmühl, Germany.

2004   The Orthographic Distinction between Velar and Glottal Spirants in Maya Hieroglyphic Writing. In *The Linguistics of Maya Writing*, ed. Soren Wichman, pp. 61–82. University of Utah Press, Salt Lake City.

Grube, Nikolai, Alfonso Lacadena, and Simon Martin
2003   Chichén Itzá and Ek Balam: Terminal Classic Inscriptions from Yucatán. In *Notebook for the XXVIIth Maya Hieroglyphic Forum at Texas*. University of Texas, Austin.

Hammond, Norman
1978   The Myth of the Milpa: Agricultural Expansion in the Maya Lowlands. In *Pre-Hispanic Maya Agriculture*, ed. Peter D. Harrison and B. L. Turner II, pp. 23–34. University of New Mexico Press, Albuquerque.

Hanks, William F.
1990   *Referential Practice: Language and Lived Space Among the Maya*. University of Chicago Press, Chicago.

Hansen, Richard Duane
1992   *The Archaeology of Ideology: A Study of Maya Preclassic Architectural Sculpture and Nakbé, Petén, Guatemala*. Ph.D. dissertation, Department of Anthropology, University of California, Los Angeles.

Harrison, Peter D.
1990   The Revolution in Ancient Maya Subsistence. In *Visions and Revision in Maya Studies*, ed. S. Clancy and Peter D. Harrison, pp. 99–113. University of New Mexico Press, Albuquerque.

1996   Settlement and Land Use in the Pulltrouser Swamp Archaeological Zone, Northern Belize. In *The Managed Mosaic: Ancient Maya Agriculture and Resource Use*, ed. Scott L. Fedick, pp. 177–92. University of Utah Press, Salt Lake City.

1999   *The Lords of Tikal: Rulers of an Ancient Maya City*. Thames and Hudson, London.

Harrison, Peter D., and B. L. Turner II (eds.)
1978   *Pre-Hispanic Maya Agriculture*. University of New Mexico Press, Albuquerque.

Hassig, Ross
1992   *War and Society in Ancient Mesoamerica*. University of California Press, Berkeley.

Haviland, William
1988   Musical Hammocks at Tikal: Problems with Reconstructing Household Composition. In *Household and Community in the Mesoamerican Past*, ed. Richard R. Wilk and Wendy A. Ashmore, pp. 121–34. University of New Mexico Press, Albuquerque.

Hayden, Brian D., and Aubrey Cannon
1982   The Corporate Group as an Archaeological Unit. *Journal of Anthropological Archaeology* 1:132–58.

Hayden, Brian D., and Huguette Sansonnet-Hayden
2001   Cognata, Capta, and Data: Hunting for Meaning. *The SAA Archaeological Record* 1(3):34–36.

Healan, Dan M.
1993   Local versus Non-local Obsidian Exchange at Tula and Its Implications for Post-Formative Mesoamerica. *World Archaeology* 24:449–66.

Hendon, Julia A.
  1992   The Interpretation of Survey Data: Two Case Studies from the Maya Area. *Latin American Antiquity* 3(1):22–42.
Hernández H., Concepción
  n.d.   La cerámica del Periodo Preclasico Tardio (300 A.C.–350 D.C.) en el norte de la Peninsula de Yucatán, México. In *La Produccion Alfarera en el México Antiguo*, ed. Norberto Gonzalez Crespo and Angel Garcia Cook. Instituto Nacional de Antropología e Historia, Mexico City.
Herrera Castro, Natividad D.
  1994   Los huertos familiares Mayas en el oriente de Yucatán. *Etnoflora Yucatánense*, vol. 9. Universidad Autónoma de Yucatán, Mérida, Yucatán, Mexico.
Herrera Castro, Natividad D., Arturo Gómez-Pompa, L. Cruz Kuri,
and Juan Salvador Flores
  1993   Los huertos familiares Mayas en X-uilub, Yucatán, México. Aspectos generales y estudio comparativo entre la flora de los huertos familiares y la selva. *Biotica Nueva Epoca* 1:19–36.
Hervik, Peter
  1999a *Mayan People within and beyond Boundaries: Social Categories and Lived Identity in Yucatán*. Harwood Academic Press, Amsterdam.
  1999b The Mysterious Maya of National Geographic. *Journal of Latin American Anthropology* 4(1):166–97.
Heyden, Doris
  1975   An Interpretation of the Cave Underneath the Pyramid of the Sun in Teotihuacan, Mexico. *American Antiquity* 40:131–47.
Hicks, Frederic
  1991   Gift and Tribute: Relations of Dependency in Aztec Mexico. In *Early State Economics*, ed. Henri J. M. Claessen and Pieter van de Velde, pp. 199–213. Transaction Publishers, New Brunswick, NJ.
Hiraoka, M.
  1986   Zonation of Mestizo Riverine Farming Systems in Northeast Peru. *National Geographic Research* (2):354–71.
Hirth, Kenneth
  1993   Identifying Rank and Socioeconomic Status in Domestic Contexts: An Example from Central Mexico. In *Prehispanic Domestic Units in Western Mesoamerica*, ed. Robert Santley and Kenneth Hirth, pp. 121–46. CRC Press, Boca Raton, FL.
Holmes, William Henry
  1895   *Archaeological Studies among the Ancient Cities of Mexico: Part I, Monuments of Yucatán*. Anthropological Series 1(1). Field Columbian Museum, Chicago.
Houck, Charles W.
  1994   The Ek Balam Rural Survey. In *Proyecto Ek Balam: Preliminary Report on the 1994 Field Season*, ed. William M. Ringle and George J. Bey. Report submitted to the National Science Foundation, Washington, DC and to the Instituto Nacional de Antropología e Historia, Mexico City.
  1996   *Rural Survey at Ek Balam, Yucatán, México: 1987–1995*. Paper presented at the 63d Annual Meeting of the Society for American Archaeology, New Orleans.
  1998a Rural Zone Reconnaissance and Excavations, 1995. In *Proyecto Ek Balam: 1995 and 1997 Field Seasons*, ed. William M. Ringle and George J. Bey III. Report pre-

pared for the National Science Foundation, Washington DC in partial fulfillment of Grant SBR-9321603.

1998b Rural Zone Reconnaissance and Excavations, 1997. In *Proyecto Ek Balam: 1995
and 1997 Field Seasons*, ed. William M. Ringle and George J. Bey III. Report prepared for the National Science Foundation, Washington, DC in partial fulfillment of Grant SBR-9321603.

1998c *Settlement and Sociopolitical Dynamics at Ek Balam: The View from the Hinterland.*
Paper presented at the 97th Annual Meeting of the American Anthropological Association, Philadelphia.

2004   *The Rural Survey of Ek Balam, Yucatán, Mexico.* Ph.D. dissertation, Department
of Anthropology, Tulane University, New Orleans.

Houston, Stephen D.
1992   *Hieroglyphs and History at Dos Pilas: Dynastic Politics of the Classic Maya.* University of Texas Press, Austin.

Hutson, Scott R., Aline Magnoni, and Bruce H. Dahlin
2000   *Intra-site Settlement Patterns at Chunchucmil, Yucatán, Mexico: A Preliminary Assessment.* Paper presented at the 65th Annual Meeting of the Society for American Archaeology, Philadelphia.

Hutson, Scott R., Aline Magnoni, and Daniel Mazeau
2001   Mapping the Urban Landscape. In *Pakbeh Regional Economy Program: Report
of the 2001 Season*, ed. B. Dahlin and Daniel Mazeau, pp. 1–2. Department of
Sociology and Anthropology, Howard University, Washington, DC.

Hutson, Scott R., Aline Magnoni, and Travis W. Stanton
2004   House Rules?: The Practice of Social Organization in Classic Period Chunchucmil, Yucatán, Mexico. *Ancient Mesoamerica* 15(1):75–92.

Jiménez Alvarez, Socorro del Pilar
2002   *La cronología cerámica del Puerto Maya de Xcambó, Costa Norte de Yucatán: Complejo cerámica Xcambó y complejo cerámico Cayalac.* Tesis de licenciatura en ciencias antropológicas, Universidad Autónoma de Yucatán, Mérida, Yucatán,
Mexico.

Johnson, A.
1983   Machiguenga Gardens. In *Adaptive Responses of Native Amazonians*, ed. R. B.
Hames and W. T. Vickers, pp. 29–64. Academic Press, NY.

Johnston, K. J., and Nancy Gonlin
1998   What Do Houses Mean? Approaches to the Analysis of Classic Maya Commoner Residence. In *Function and Meaning in Classic Maya Architecture*, ed.
Stephen D. Houston, pp. 141–85. Dumbarton Oaks Research Library and Collection, Washington, DC.

Johnstone, Dave
1994   Excavations Within the Ballcourt Plaza. In *The Selz Foundation Yaxuná Project:
Report of the 1993 Field Season*, ed. C. Suhler, pp. 62–69. Department of Anthropology, Southern Methodist University, Dallas.

2001   *Ceramics of Yaxuná, Yucatán.* Ph.D. dissertation, Southern Methodist University, Dallas.

Jones, Lindsay
1995   *Twin City Tales: A Hermeneutical Reassessment of Tula and Chichén Itzá.* University Press of Colorado, Niwot.

Justeson, John S., William M. Norman, Lyle Campbell, and Terrence Kaufman
  1985   *The Foreign Impact on Lowland Maya Language and Script.* Middle American Research Institute Publication 53. Tulane University, New Orleans.
Karmack, A. M.
  1962   *Climate and Economic Development.* EDI Seminar Paper no. 2. Economic Development Institute, International Bank for Reconstruction and Development, Washington, DC.
Kelley, David H.
  1992   Yucatán y el Imperio Tolteca. *Arqueología* 8:113–19.
Kepecs, Susan
  1997   Native Yucatán and Spanish Influence: The Archaeology and History of Chikinchel. *Journal of Archaeological Method and Theory* 4(3–4):307–29.
  1998   Diachronic Ceramic Evidence and Its Social Implications in the Chikinchel Region, Northeast Yucatán, Mexico. *Ancient Mesoamerica* 9:121–36.
  1999   *The Political Economy of Chikinchel, Yucatán, Mexico: A Diachronic Analysis from the Prehispanic Era through the Age of Spanish Administration.* Ph.D. dissertation, Department of Anthropology, University of Wisconsin–Madison. University Microfilms, Ann Arbor, MI.
Kepecs, Susan, and Sylviane Boucher
  1996   The Pre-Hispanic Cultivation of Rejolladas and Stone-Lands: New Evidence from Northeast Yucatán. In *The Managed Mosaic: Ancient Maya Agriculture and Resource Use*, ed. Scott L. Fedick, pp. 69–91. University of Utah Press, Salt Lake City.
Kepecs, Susan M., Gary M. Feinman, and Sylviane Boucher
  1994   Chichén Itzá and Its Hinterland: A World-Systems Perspective. *Ancient Mesoamerica* 5:141–58.
Kerr, Justin
  1989   *The Maya Vase Book Volume 1.* Kerr Associates, NY.
  1992   *The Maya Vase Book Volume 3.* Kerr Associates, NY.
Killion, Thomas W.
  1990   Cultivation Intensity and Residential Site Structure: An Ethnoarchaeological Examination of Peasant Agriculture in the Sierra de los Tuxtlas, Veracruz, México. *Latin American Antiquity* 1(3):191–213.
  1992   Residential Ethnoarchaeology and Ancient Site Structure: Contemporary Farming and Prehistoric Settlement Agriculture at Matacapan, Veracruz, Mexico. In *Gardens of Prehistory: The Archaeology of Settlement Agriculture in Greater Mesoamerica*, ed. Thomas W. Killion, pp. 119–49. University of Alabama Press, Tuscaloosa.
Kimber, C. T.
  1973   Spatial Patterning in Dooryard Gardens of Puerto Rico. *Geographical Review* (63):6–26.
Kowalewski, Stephen A., Richard E. Blanton, Gary M. Feinman, and Laura Finsten
  1983   Boundaries, Scale, and Internal Organization. *Journal of Anthropological Archaeology* 2:32–56.
Kowalski, Jeff K.
  1985   Lords of the Northern Maya: Dynastic History in the Inscriptions of Uxmal and Chichén Itzá. *Expedition* 27(3):50–60.
  1987   *The House of the Governor: A Maya Palace at Uxmal, Yucatán, Mexico.* University of Oklahoma, Norman.

Kowalski, Jeff K., Alfredo Barrera R., Hever Ojeda M., and José Huchim H.
    1996   Archaeological Excavations of a Round Temple at Uxmal: Summary Discussion and Implications for Northern Maya Culture History. In *Eighth Palenque Round Table, 1993*, ed. Martha J. Macri and Jan McHargue, pp. 281–96. Pre-Columbian Art Research Institute, San Francisco.
Kramer, Carol
    1979   An Archaeological View of a Contemporary Kurdish Village: Domestic Architecture, Household Size, and Wealth. In *Ethnoarchaeology: Implications of Ethnography for Archaeology*, ed. Carol Kramer, 139–63. Columbia University Press, NY.
Kristan-Graham, Cynthia
    1992   The Business of Narrative at Tula: An Analysis of the Vestibule Frieze, Trade, and Ritual. *Latin American Antiquity* 4:3–21.
Krochock, Ruth J.
    1988   *The Hieroglyphic Inscriptions and Iconography of the Temple of the Four Lintels and Related Monuments, Chichén Itzá, Yucatán, Mexico*. Master's thesis, University of Texas, Austin.
    1998   *The Development of Political Rhetoric at Chichén Itzá, Yucatán, Mexico*. Ph.D. dissertation, Department of Anthropology, Southern Methodist University, Dallas.
Kubler, George
    1961   Chichén-Itzá y Tula. *Estudios de Cultura Maya* 1:47–79.
Kunen, Julie L.
    2004   *Ancient Maya Life in the Far West Bajo: Social and Environmental Change in the Wetlands of Belize*. University of Arizona Press, Tucson.
Kunen, Julie L., T. Patrick Culbert, Vilma Fialdo, Brian R. McKee, and Liwy Grazioso
    2000   *Bajo* Communities: A Case Study from the Central Peten. *Culture and Agriculture* 22(3):15–31.
Kurjack, Edward B.
    1974   *Prehistoric Lowland Maya Community and Social Organization: A Case Study at Dzibilchaltún, Yucatán, Mexico*. Middle American Research Institute Publication 38. Tulane University, New Orleans.
    1977   Sacbeob: Parentesco y Desarrollo del Estado. *XV Mesa Redonda* 1:217–30.
    1979   *Introduction to the Map of the Ruins of Dzibilchaltún, Yucatán, Mexico*. Middle American Research Institute Publication 47. Tulane University, New Orleans.
    1992   Conflicto en el arte de Chichén Itzá. *MAYAB* 8:88–96.
    2003   Sitios, monumentos y la organización territorial de los Mayas Precolombinos. In *Los Investigadores de la Cultura Maya 11*, 1:9–18. Universidad Autónoma de Campeche, Campeche, Mexico.
Kurjack, Edward B., and E. Wyllys Andrews V
    1976   Early Boundary Maintenance in Northwest Yucatán, Mexico. *American Antiquity* 41(3):318–25.
Lanfant, M. F.
    1995   International Tourism, Internationalization and the Challenge to Identity. *Sage Studies in International Sociology* 47:24–43.
Lentz, David L.
    1999   Plant Resources of the Ancient Maya: The Paleoethnobotanical Evidence. In *Reconstructing Ancient Maya Diet*, ed. Christine D. White, pp. 3–18. University of Utah Press, Salt Lake City.

Lesser, Juan M., and A. E. Weidie
    1988   Region 25, Yucatán Peninsula. In *Hydrogeology*, ed. William Back, Joseph S.
            Rosenshein, and Paul R. Seaber, pp. 237–41. The Geology of North America
            vol. O-2. The Geological Society of America, Boulder, CO.
Lightfoot, Kent G., and Antoinette Martinez
    1995   Frontiers and Boundaries in Archaeological Perspective. *Annual Review of An-
            thropology* 24:471–92.
Lincoln, Charles E.
    1980   *A Preliminary Assessment of Izamal, Yucatán, Mexico*. Undergraduate thesis, Tu-
            lane University, New Orleans.
    1985   Ceramics and Ceramic Chronology. In *A Consideration of the Early Classic Pe-
            riod in the Maya Lowlands*, ed. Gordon R. Willey and Peter Mathews, pp. 55–
            94. Institute of Mesoamerican Studies, Publication 10. State University of New
            York at Albany.
    1986   The Chronology of Chichén Itzá: A Review of the Literature. In *Late Lowland
            Maya Civilization*, ed. Jeremy A. Sabloff and E. Wyllys Andrew V, pp. 141–96.
            University of New Mexico Press, Albuquerque.
    1990   *Ethnicity and Social Organization at Chichén Itzá, Yucatán, Mexico*. Ph.D. disser-
            tation, Department of Anthropology, Harvard University, Cambridge, MA.
Littman, Edwin R.
    1958   Ancient Mesoamerican Mortars, Plasters, and Stuccos: The Composition and
            Origin of Sascab. *American Antiquity* 24:172–76.
Lizardi Ramos, Cesar
    1940   *Exploraciones en Quintana Roo*. Mexico City.
Lohse, Jon C., and Patrick N. Findlay
    2000   A Classic Maya House-Lot Drainage System in Northwestern Belize. *Latin
            American Antiquity* 11:175–85.
Lohse, Jon C., and Dale Hudler
    1997   *Measuring Social Diversity among Ancient Maya Households*. Paper presented at
            the 62d Annual Meeting of the Society for American Archaeology, Nashville.
Lorenzen, Karl James
    1999   New Discoveries at Tumben-Naranjal: Late Postclassic Reuse and the Ritual
            Recycling of Cultural Geography. *Mexicon* 21:98–107.
Lundell, Cyrus L.
    1933   The Agriculture of the Maya. *Southwest Review* 19:65–77.
Lynott, Mark J., and Alison Wylie (eds.)
    2000   *Ethics in American Archaeology*. The Society for American Archaeology, Wash-
            ington, DC.
Magnoni, Aline, Scott Hutson, and Timothy Beach
    2002   *Chemical Analyses for the Identification of Activity Areas in Two Residential Groups
            at Chunchucmil, Yucatán*. Paper presented at the 67th Annual Meeting of the
            Society for American Archaeology, Denver.
Magnoni, Aline, Scott R. Hutson, and Tara M. Bond-Freeman
    2000   *The Peculiar Cultural Landscape of Chunchucmil*. Paper presented at the 99th
            Annual Meeting of the American Anthropological Association, San Francisco.
Magnoni, Aline, Scott R. Hutson, and Travis W. Stanton
    2002   *Urban Landscape Transformation and Perceptions at Chunchucmil, Yucatán*. Paper
            presented at the 101st Annual Meeting of the American Anthropological As-
            sociation, New Orleans.

Maldonado, Rubén

1980   *Informe de trabajo de campo realizado por el Proyecto Aké durante la temporada de 1979*. Report prepared for the Sección de Arqueología del Centro, Instituto Nacional de Historia e Antropología. Mérida, Yucatán, Mexico.

1981   *Informe del trabajo de campo realizado por el Proyecto Aké durante la temporada de 1981*. Report prepared for the Instituto Nacional de Historia e Antropología, Mexico City.

1989   Restauración del edificio de las pilastras de Aké, Yucatán. *Revista Española de Antropología Americana*, 19:27–48. Facultad de Geografía e Historia, Universidad Complutense de Madrid, Madrid.

1990   Izamal, sitio hegemónico del norte de Yucatán en la Epoca Clásica. In *La Epoca Clásica: Nuevos Hallazgos, Nuevas Ideas*, ed. Amalia Cardo de Mendez, pp. 483–96. Museo Nacional de Antropología, Instituto Nacional de Antropología e Historia, Mexico City.

1995   Los sistemas de caminos del norte de Yucatán. In *Seis Ensayos sobre Antiguos Patrones de Asentamiento en el Área Maya*, pp. 68–92. Instituto de Investigaciones Antropológicas, Universidad Nacional Autónoma de México, Mexico City.

Maldonado, Rubén, and Edward B. Kurjack

1993   Reflexiones sobre las relaciones entre Chichén Itzá, sus vecinos peninsulares y Tula. *Arqueología* 2a. época., 9–10:97–103. Instituto Nacional de Antropología e Historia, Mexico City.

Maldonado, Rubén, Alexander Voss, and Ángel Góngora

2001   Kalom Uk'Uw, Señor de Dzibilchaltún. La organización social entre los Mayas. *Memoria de la Tercera Mesa Redonda de Palenque*, 1:79–100. Instituto Nacional de Historia e Antropología and Autonomous University of Yucatan, Mexico.

Manahan, Kam, James N. Ambrosino, and Traci A. Ardren

1997   *The Last Stand: Defensive Constructions at Yaxuná, Yucatán*. Paper presented at the 62d Annual Meeting of the Society for American Archaeology, Nashville.

Mansell, Eugenia B., and Tara M. Bond-Freeman

2002   The Preliminary Ceramic Report for the 2001 Field Season. In *Pakbeh Regional Economy Program: Informe 2001 Field Season*, ed. Bruce H. Dahlin and Daniel E. Mazeau, pp. 114–222. Department of Sociology and Anthropology, Howard University, Washington, DC.

Manzanilla, Linda

1987   *Cobá, Quintana Roo: Analisis de dos unidades habitacionales Mayas*. Universidad Nacional Autónoma de México, Mexico City.

1996   Corporate Groups and Domestic Activities at Teotihuacan. *Latin American Antiquity* 7:228–46.

Marcus, Joyce

1973   Territorial Organization of the Lowland Classic Maya. *Science* 180:911–16.

1983   On the Nature of the Mesoamerican City. In *Prehistoric Settlement Patterns: Essays in Honor of Gordon R. Willey*, ed. Evan Z. Vogt and Richard M. Leventhal, pp. 195–242. University of New Mexico, Albuquerque; Peabody Museum, Harvard University, Cambridge, MA.

1984   Mesoamerican Territorial Boundaries: Reconstructions from Archaeology and Hieroglyphic Writing. *Archaeological Review from Cambridge* 3(2):48–62.

1993   Ancient Maya Political Organization. In *Lowland Maya Civilization in the Eighth Century AD*, ed. Jeremy A. Sabloff and John S. Henderson, pp. 111–83. Dumbarton Oaks Research Library and Collection, Washington, DC.

1995   Where Is Lowland Maya Archaeology Headed? *Journal of Archaeological Research* 3:3–53.

2003   The Maya and Teotihuacan. In *The Maya and Teotihuacan, Reinterpreting Early Classic Interaction*, ed. Geoffrey E. Braswell, pp. 337–56. University of Texas Press, Austin.

Marshall, Yvonne

2002   What Is Community Archaeology? *World Archaeology* 34(2):211–19.

Martin, Simon

1997   The Painted King List: A Commentary on Codex-Style Dynastic Vases. In *The Maya Vase Book, a Corpus of Rollout Photographs of Maya Vases*, ed. Justin Kerr, Vol. 5, pp. 846–67. Kerr Associates, NY.

Martin, Simon, and Nikolai Grube

1995   Maya Superstates. *Archaeology* 48(6):41–46.

2000   *Chronicle of the Maya Kings and Queens*. Thames and Hudson, London.

Mastache, Alba Guadalupe, Robert H. Cobean, and Dan M. Healan

2002   *Ancient Tollan, Tula and the Toltec Heartland*. University Press of Colorado, Boulder.

Mathews, Jennifer P.

1995   The Box Ni Group of Naranjal, Quintana Roo, and Early Architecture of the Central Maya Lowlands. In *The View From Yalahau: Archaeological Investigations in Northern Quintana Roo, Mexico*, pp. 79–87. Latin American Studies Program, Field Report Series, no. 2. University of California Press, Riverside.

1998   *The Ties that Bind: The Ancient Maya Interaction Spheres of the Late Preclassic and Early Classic Periods in the Northern Yucatán Peninsula*. Ph.D. dissertation, Department of Anthropology, University of California, Riverside. University Microfilms, Ann Arbor, MI.

2001   Radiocarbon Dating of Mortar and Charcoal Inclusions in Architectural Mortar: A Case Study in the Maya Region, Quintana Roo, Mexico. *Journal of Field Archaeology* 28(3–4):395–400.

2003   Megalithic Architecture at the Site of Victoria, Quintana Roo. *Mexicon* XXV (3):74–77.

Mathews, Jennifer P., and Scott L. Fedick

2000   *T'isil: A Late Preclassic Maya Community in the Yalahau Region of Quintana Roo*. Paper presented at the 65th Annual Meeting of the Society for American Archaeology, Philadelphia.

Mathews, Peter

1991   Classic Maya Emblem Glyphs. In *Classic Maya Political History*, ed. T. P. Culbert, pp. 19–29. Cambridge University Press, Cambridge.

Mazeau, Daniel E.

2001   Analysis of Lithic Material at Chunchucmil. In *Pakbeh Regional Economy Program: Report of the 2001 Field Season*, ed. Bruce H. Dahlin and Daniel E. Mazeau, pp. 126–33. Sociology and Anthropology Department, Howard University, Washington, DC.

Mazeau, Daniel E., and Jamie Forde

2003   *The Lithic Industries of Chunchucmil: Chipped Stone Tools in a Market Economy*. Paper presented at the 68th Annual Meeting of the Society for American Archaeology, Milwaukee, WI.

McAnany, Patricia A.

1992   Agricultural Tasks and Tools: Patterns of Stone Tool Discard near Prehistoric

Maya Residences Bordering Pulltrouser Swamp, Belize. In *Gardens of Prehistory: The Archaeology of Settlement Agriculture in Greater Mesoamerica*, ed. Tom Killion, pp. 184–213. University of Alabama, Tuscaloosa.

McCall, John
1995   *Preliminary Report of 1995 Survey Results*. Manuscript in possession of Charles Houck.

McNeil, Mary
1964   *Spatial and Temporal Patterns of Prehistoric Settlement, Procurement, and Exchange on the Coast and Cays of Southern Belize*. Ph.D. dissertation, University of Wisconsin, Madison.

Michelet, Dominique, Pierre Becquelin, and Marie-C. Arnauld
2000   *Mayas del Puuc: Arqueologia de la Region de Xculoc, Campeche*. Gobierno del Estado de Campeche, Centre d'Etudes Mexicaines et Centroaméricaines, Campeche, Mexico.

Milbrath, Susan, and Carlos Peraza Lope
2003   Revisiting Mayapán: México's Last Maya Capital. *Ancient Mesoamerica* 14:1–46.

Miller, Arthur G.
1973   *The Mural Painting of Teotihuacan*. Dumbarton Oaks Research Library and Collection, Washington, DC.

Millet, Luis, and Rafael Burgos Villanueva
1998   *Proyecto Izamal. Avances y perspectivas*. Paper presented at El Congreso Internacional de Mayistas, Antigua, Guatemala.

Moholoy-Nagy, Hattula
1999   Mexican Obsidian at Tikal, Guatemala. *Latin American Antiquity* 10:300–313.

Molina Montes, Augusto
1982   Archaeological Buildings: Restoration or Misrepresentation. In *Falsifications and Misreconstructions of Pre-Columbian Art*, ed. Elizabeth H. Boone, pp. 125–41. Dumbarton Oaks Research Library and Collection, Washington, DC.

Montejo, Victor
1997   *A Maya Reflects on 500 Years: Cortez to the Present*. Paper presented at the 4th Annual UCLA Maya Weekend, Los Angeles.

Morley, Sylvanus G.
1919   *Archaeology*. Carnegie Institution of Washington, Yearbook No. 17 (1918):269–76.
1946   *The Ancient Maya*. Stanford University Press, Palo Alto, CA.

Morrison, Bethany A.
2000   *Ancient Maya Settlement of the Yalahau Region: An Example from the El Edén Wetland*. Ph.D. diss., Dept. of Anthropology, Univ. of California, Riverside.

Morrison, Bethany A., and Roberto Cózatl Manzano
2003   Initial Evidence for Use of Periphyton as an Agricultural Fertilizer by the Ancient Maya Associated with the El Edén Wetland, Northern Quintana Roo, Mexico. In *The Lowland Maya Area: Three Millennia at the Human-Wildland Interface*, ed. Arturo Gomez-Pompa, Michael F. Allen, Scott L. Fedick, and Juan J. Jiménez-Osornio. Haworth Press, Binghamton, NY.

Munt, I.
1994   The "Other" Post-Modern Tourism: Culture, Travel and the New Middle Classes. *Theory, Culture and Society* 11:101–23.

Narroll, Raoul
   1962   Floor Area and Settlement Population. *American Antiquity* 27:587–89.
Nash, D.
   1989   Tourism as a Forum of Imperialism. In *Hosts and Guests: The Anthropology of Tourism*, ed. Valene L. Smith, 2d ed., pp. 37–47. University of Pennsylvania Press, Philadelphia.
National Center for Environmental Assessment
   1999   *Karst Terminology with Special Reference to Environmental Karst Hydrology*. National Center for Environmental Assessment–Washington Division, Office of Research and Development, U. S. Environmental Protection Agency, Washington, DC.
Navarrete, Carlos, María José Con Uribe, and Alejandro Martínez Muriel
   1979   *Observaciones arqueológicas en Cobá, Quintana Roo*. Universidad Nacional Autónoma de México, Mexico City.
Neff, Hector
   2002   Nuevos hallazgos relacionados con la producción de la Vajilla Plomiza. In *XV Simposio de Investigaciones Arqueológicas en Guatemala, 2001*, ed. Juan Pedro Laporte, Héctor Escobedo, and Bárbara Arroyo, 2:529–42. Museo Nacional de Arqueología y Etnología, Guatemala.
Ortega, Luz Maria, Sergio Avendano, Arturo Gomez-Pampa, and Edilberto Ucan Ek
   1993   Los solares de Chunchucmil, Yucatán, México. *Biotica Nueva Epoca* 1:37–51.
Ower, L.
   1927   Features of British Honduras. *Geographical Journal* 70:372–86.
Pacheco Benítez, Adán, and Ana M. Parrilla Albuerne
   2004   El Naranajal, Quintana Roo (México). Un sitio Megalítico en las tierras bajas del norte. *MAYAB* 17:5–19.
Palacios-Mayorga, Sergio, Ana Luisa Anaya, Eleazar González-Velázquez,
Lázaro Huerta-Arcos, and Arturo Gómez-Pompa
   2003   Periphyton as a Potential Biofertilizer in Intensive Agriculture of the Ancient Maya. In *The Lowland Maya Area: Three Millennia at the Human-Wildland Interface*, ed. Arturo Gomez-Pompa, Michael F. Allen, Scott L. Fedick, and Juan J. Jiménez-Osornio. Haworth Press, Binghamton, NY.
Palerm, Angel, and Eric R. Wolf
   1957   Ecological Potential and Cultural Development in Mesoamerica. In *Studies in Human Ecology*. Anthropological Society of Washington and Pan American Union Social Science Monograph no. 3. Pan American Union, Washington, DC.
Pasztory, Esther
   1993   Catalogue of Objects. In *Teotihuacan: Art from the City of the Gods*, ed. K. Berrin and E. Pasztory, p. 204. Thames and Hudson, NY.
Patch, Robert W.
   1993   *Maya and Spaniard in Yucatan, 1648–1812*. Stanford University Press, Palo Alto, CA.
Peraza Lope, Carlos
   1999   Mayapán: Ciudad-capital del Postclásico. *Arqueología Mexicana* VII(37):48–53.
Peraza Lope, Carlos, Pedro Delgado, and Barbara Escamilla
   2002   Intervenciones en un edificio del Preclásico medio en Tipikal, Yucatán. In *Los Investigadores de la Cultura Maya*, 10, 1:263–276. Universidad Autónoma de Campeche, Campeche, Mexico.

Pérez de Heredia Puente, Eduardo
   1998   Datos recientes sobre la cerámica de Chichén Itzá. In *Los Investigadores de la Cultura Maya 6*, 2:271–87. Universidad Autónoma de Campeche, Campeche, Mexico.

Pohl, Mary D. (ed.)
   1990   *Ancient Maya Wetland Agriculture: Excavations on Albion Island, Northern Belize.* Westview Press, Boulder, CO.

Pokotylo, David, and Neil Guppy
   1999   Public Opinion and Archaeological Heritage: Views from Outside the Profession. *American Antiquity* 64(3):400–416.

Pollock, Harry E. D.
   1980   *The Puuc: An Architectural Survey of the Hill Country of Yucatán and Northern Campeche, Mexico.* Peabody Museum of Archaeology and Ethnology Memoirs 19. Harvard University, Cambridge, MA.

Pollock, Harry E. D., Ralph L. Roys, Tatiana Proskouriakoff, and A. Ledyard Smith
   1962   *Mayapan, Yucatan, Mexico.* Publication 619. Carnegie Institution of Washington, Washington, DC.

Proskouriakoff, Tatiana
   1950   *A Study of Classic Maya Sculpture.* Publication 593. Carnegie Institution of Washington, Washington, DC.
   1970   On Two Inscriptions at Chichén Itzá. In *Monographs and Papers in Maya Archaeology*, ed. William R. Bullard, pp. 457–67. Papers of the Peabody Museum, Vol. 61. Peabody Museum, Cambridge, MA.

Puleston, Dennis E.
   1968   *Brosimum alicastrum as a Subsistence Alternative for the Classic Maya of the Central Southern Lowlands.* Master's thesis, University of Pennsylvania.
   1983   *The Settlement Survey of Tikal.* Tikal Reports No. 13. University Museum Monograph 48. University of Pennsylvania, Philadelphia.

Pyburn, K. Anne
   1989   *Prehistoric Maya Community and Settlement at Nohmul, Belize.* BAR International Series, No. 509. British Archaeological Reports, Oxford.
   1996   The Political Economy of Ancient Maya Land Use: The Road to Ruin. In *The Managed Mosaic: Ancient Maya Agriculture and Resource Use*, ed. Scott L. Fedick, pp. 236–50. University of Utah Press, Salt Lake City.
   1997   The Archaeological Signature of Complexity in the Maya Lowlands. In *The Archaeology of City-States: Cross-Cultural Approaches*, ed. Deborah L. Nichols and Thomas H. Charlton, pp. 155–68. Smithsonian Institution Press, Washington, DC.
   1998   Smallholders in the Maya Lowlands: Homage to a Garden Variety Ethnographer. *Human Ecology* 26:267–86.
   2003   We Have Never Been Post-modern: Maya Archaeology in the Ethnographic Present. In *Continuities and Changes in Maya Archaeology: Perspectives at the Millennium*, ed. Charles W. Golden and Greg Borgstede, pp. 287–93. Routledge, NY.

Pyburn, K. Anne, and Richard R. Wilk
   2000   Responsible Archaeology Is Applied Anthropology. In *Ethics in American Archaeology*, ed. Mark J. Lynott and Alison Wylie, pp. 78–83. Society for American Archaeology, Washington, DC.

Quezada, Sergio
  1993   *Pueblos y caciques Yucatecos, 1550–1580.* El Colegio de México, Mexico City.
Quintal Suaste, Beatriz
  1993   *Los asentamientos arqueológicos de Aké, Yucatán: Una aproximación a su organización social.* Undergraduate thesis, Facultad de Ciencias Antropológicas de la Universidad Autónoma de Yucatán, Mérida, Yucatán, Mexico.
Rathje, William L.
  1983   To the Salt of the Earth. In *Prehistoric Settlement Patterns: Essays in Honor of Gordon R. Willey,* ed. E. Vogt and Richard Leventhal, pp. 23–34. Peabody Museum of Archaeology and Ethnology, Harvard University, Cambridge, MA.
Rattray, Evelyn C.
  1987   Los barrios foráneos de Teotihuacan. In *Teotihuacan: Nuevos Datos, Nuevas Síntesis, Nuevos Problemas,* ed. Emily McClung de Tapia and Evelyn Childs Rattray, pp. 243–73. Universidad Nacional Autonoma de México, Mexico City.
  2001   *Teotihuacan: Ceramics, Chronology, and Cultural Trends.* Instituto Nacional de Antropología e Historia/University of Pittsburgh, Mexico City.
Re Cruz, Alicia
  1996a   The Thousand and One Faces of Cancun. *Urban Anthropology* 25(3):283–310.
  1996b   *The Two Milpas of Chan Kom: Scenarios of a Maya Village Life.* State University of New York Press, Albany.
  2003   Milpa as an Ideological Weapon: Tourism and Maya Migration to Cancún. *Ethnohistory* 50(3):489–502.
Redfield, Robert
  1941   *The Folk Culture of Yucatán.* University of Chicago Press, Chicago.
  1950   *A Village that Chose Progress: Chan Kom Revisited.* University of Chicago Press, Chicago.
Redfield, Robert, and Alfonso Villa Rojas
  1934   *Chan Kom: A Maya Village.* Publication 448. Carnegie Institution of Washington, Washington, DC.
Redmond, Elsa
  1983   *A fuego y sangre: Early Zapotec Imperialism in the Cuicatlán Cañada, Oaxaca.* Memoirs No. 16. Museum of Anthropology, University of Michigan, Ann Arbor, MI.
Reents-Budet, Dorie
  1998   Elite Maya Pottery and Artisans as Social Indicators. In *Craft and Social Identity,* ed. Cathy L. Costin and Rita P. Wright, pp. 71–89. Archaeological Papers of the American Anthropological Association. American Anthropological Association, Arlington, VA.
Repetto Tió, Beatriz
  1986   *Informe preliminar del trabajo de campo del Proyecto Dzibilchaltún-Sacbé 1.* Manuscript on file with the Archivo del Centro, Instituto Nacional de Antropología e Historia, Mérida, Yucatán, Mexico.
Restall, Matthew
  1997   *The Maya World: Yucatec Culture and Society, 1550–1850.* Stanford University Press, Palo Alto, CA.
Rice, Don S., and Prudence M. Rice
  1990   Population Size and Population Change in the Central Petén Lakes Region, Guatemala. In *Precolumbian Population History in the Maya Lowlands,* ed.

248 Bibliography

T. Patrick Culbert and Don S. Rice, pp. 123–48. University of New Mexico Press, Albuquerque.

Rice, Prudence M.
1987   Economic Change in the Lowland Maya Late Classic Period. In *Specialization, Exchange and Complex Societies*, ed. E. M. Brumfiel and T. K. Earle, pp. 76–85. Cambridge University Press, Cambridge.

Ricketson, Oliver G., and Edith Bayles Ricketson
1937   *Uaxactun, Guatemala. Group E–1926–1931*. Publication 477. Carnegie Institution of Washington, Washington, DC.

Rider, Nick
1999   *Yucatán & Southern Mexico*. Cambridge University Press, Cambridge.

Ringle, William M.
1990   Who Was Who in Ninth-Century Chichén Itzá? *Ancient Mesoamerica* 1:233–43.
1999   Pre-Classic Cityscapes: Ritual Politics among the Early Lowland Maya. In *Social Patterns in Pre-Classic Mesoamerica*, ed. D. C. Grove and R. A. Joyce, pp. 183–223. Dumbarton Oaks Research Library and Collection, Washington, DC.
2004   On the Political Organization of Chichén Itzá. *Ancient Mesoamerica* 15:167–218.
2005   *Comments on the Symposium "Nuevas perspectivas acerca del origen y antigüedad de la vida civilizada en el norte de Yucatán"*. Paper at the 2d Congreso Internacional de Cultura Maya, Mérida, Yucatán, Mexico.

Ringle, William M., and E. Wyllys Andrews V
1988   Formative Residences at Komchen, Yucatán, Mexico. In *Household and Community in the Mesoamerican Past*, ed. Richard R. Wilk and Wendy A. Ashmore, pp. 171–97. University of New Mexico Press, Albuquerque.

Ringle, William M., and George J. Bey III
1992   *The Center and Segmentary State Dynamics: African Models in the Maya Lowlands*. Paper presented at the Segmentary State and the Classic Lowland Maya: A New Model for Ancient Political Organization conference, Cleveland State University, Cleveland.
1995   *Proyecto Ek Balam: Preliminary Report on the 1994 Field Season*. Report submitted to the National Science Foundation, Washington, DC and to the Instituto Nacional de Antropología e Historia, Mexico City.
1998   Conclusions. In *Proyecto Ek Balam: 1995 and 1997 Field Seasons*, ed. William M. Ringle and George J. Bey III. Report prepared for the National Science Foundation, Washington, DC in partial fulfillment of Grant SBR-9321603.
2001   Post-Classic and Terminal Classic Courts of the Northern Maya Lowlands. In *Royal Courts of the Ancient Maya, Volume 2: Data and Case Studies*, ed. Takeshi Inomata and Stephen D. Houston, pp. 266–307. Westview Press, Boulder, CO.

Ringle, William M., George J. Bey III, Tara M. Bond-Freeman, Craig A. Hanson, Charles W. Houck, and J. Gregory Smith
2004   The Decline of the East: The Classic to Postclassic Transition at Ek Balam, Yucatán. In *The Terminal Classic in the Maya Lowlands*, ed. Arthur A. Demarest, Prudence M. Rice, and Don S. Rice, pp. 485–516. University Press of Colorado, Boulder.

Ringle, William M., George J. Bey III, Tara M. Bond-Freeman, Charles W. Houck, Craig A. Hanson, and J. Gregory Smith
2003   El Proyecto Ek Balam: Una Perspectiva Regional, 1986–1999. *Los Investigadores*

de la Cultura Maya 11, 2:392–405. Universidad Autónoma de Campeche, Campeche, Mexico.

Ringle, William M., George J. Bey III, and Carlos Peraza Lope
  1989 *Preliminary Report of the Ek Balam Project: 1987 Field Season.* Preliminary report submitted to the National Geographic Society, Washington, DC in partial fulfillment of Award 3544–87.
  1991 *Preliminary Report of the Proyecto Ek Balam: Temporada 1989.* Submitted to the Instituto Nacional de Antropología e Historia, Mexico City, and the National Geographic Society, Washington DC.

Ringle, William M., Tomás Gallareta Negrón, and George J. Bey III
  1998 The Return of Quetzalcoatl: Evidence for the Spread of a World Religion during the Epiclassic Period. *Ancient Mesoamerica* 9:183–232.

Ringle, William M., and J. Gregory Smith
  1998 *A Report on the 1997 Field Season at Ichmul de Morley, Yucatán.* Report submitted to the National Science Foundation, Washington, DC and to the Instituto Nacional de Antropología e Historia, Mexico City. Manuscript on file at Davidson College, Davidson, NC.

Rissolo, Dominique
  1997 A Survey of Caves in the Southern Yalahau Region. In *The Yalahau Project: Preliminary Technical Report on 1996 Archaeological Investigations in Northern Quintana Roo, Mexico*, ed. Scott L. Fedick, pp. 11–22. Prepared for the Consejo de Arqueología de Instituto Nacional de Antropología e Historia, Mexico City.
  2003 *Ancient Maya Cave Use in the Yalahau Region, Northern Quintana Roo, Mexico.* Association for Mexican Cave Studies Bulletin 12. Austin, TX.

Rissolo, Dominique, and José Manuel Ochoa Rodriguez
  2002 *A Reassessment of the Middle Preclassic in Northern Quinatan Roo.* Paper presented at the 67th Annual Meeting of the Society for American Archaeology, Denver.

Rissolo, Dominique, José Manuel Ochoa Rodriguez, and Joseph Ball
  2005 A Reassessment of the Middle Preclassic in Northern Quintana Roo. In *Quintana Roo Archaeology*, ed. Justine M. Shaw and Jennifer P. Mathews. University of Arizona Press, Tucson.

Rivera Dorado, Miguel
  1987 *Oxkintok 1.* Mision Arqueológica de España en México, Madrid.
  1989 *Oxkintok 2.* Mision Arqueológica de España en México, Madrid.
  1990 *Oxkintok 3.* Mision Arqueológica de España en México, Madrid.
  1991 *Oxkintok, una Ciudad Maya de Yucatán.* Mision Arqueología de España en México, Madrid.
  1992 *Oxkintok 4.* Mision Arqueológica de España en México, Madrid.

Robertson, Merle G.
  1986 Some Observations on the X'telhu Panels at Yaxcaba, Yucatan. In *Research and Reflection in Archaeology and History: Essays in Honor of Doris Stone*, ed. E. Wyllys Andrews V, pp. 87–112. Middle American Research Institute Publication 57. Tulane University, New Orleans.

Robin, Cynthia
  2002 Outside of Houses: The Practices of Everyday Life at Chan Nòohol, Belize. *Journal of Social Archaeology* 2(2):245–67.

Robles Castellanos, Fernando
  1990   *La secuencia cerámica de la Region de Cobá, Quintana Roo.* Serie Arqueología 184.
          Instituto Nacional de Antropología e Historia, Mexico City.
  n.d.   Las esferas cerámicas Cehpech y Sotuta del apogeo del Clásico Tardío (C. 730–
          900 dc) en el norte de la Península de Yucatán. In *La Produccion Alfarera en el
          México Antiguo,* ed. Norberto Gonzalez Crespo and Angel Garcia Cook. Insti-
          tuto Nacional de Antropología e Historia, Mexico City.
  In press   El nuevo paradigma de la cronología e índole de la genesis de la civilización
          Maya en el norte de la Península de Yucatán. In *Origins of Maya Civilization.
          First Boundary End Conference on Ancient Maya,* ed. William Saturno and David
          Stuart. Center for Maya Research, Barnardsville, North Carolina.
Robles Castellanos, Fernando, and Anthony P. Andrews
  1986   A Review and Synthesis of Recent Postclassic Archaeology in Northern Yuca-
          tán. In *Late Lowland Maya Civilization: Classic to Postclassic,* ed. Jeremy A. Sab-
          loff and E. Wyllys Andrews V, pp. 53–98. University of New Mexico Press, Albu-
          querque.
  2003   *Proyecto Costa Maya: Reconocimiento Arqueológico en el Noroeste de Yucatán, Méx-
          ico.* Report for the Consejo Nacional de Arqueología de México, Yucatán,
          Mexico.
Roys, Ralph L.
  1933   *The Book of the Chilam Balam of Chumayel.* Publication 438. Carnegie Institu-
          tion of Washington, Washington, DC.
  1943   *The Indian Background of Colonial Yucatán.* Publication 548. Carnegie Institu-
          tion of Washington, Washington, DC.
  1957   *The Political Geography of Yucatán.* Publication 613. Carnegie Institution of
          Washington, Washington, DC.
Roys, Ralph L., and Edwin M. Shook
  1966   Preliminary Report on the Ruins of Aké, Yucatán. *Society for American Archae-
          ology Memoirs* 20:1–54.
Ruz L'huillier, Alberto
  1951   Chichén Itzá y Palenque, ciudades fortificadas. In *Homenaje al Doctor Alfonso
          Laso.* Imprenta Nuevo Mundo, Mexico.
  1964   Influencias mexicanas sobre los Mayas. In *Desarrollo Cultural de los Mayas,* ed.
          Evon Z. Vogt and Alberto Ruz L'huillier, pp. 195–227. Universidad Nacional
          Autónoma de México, Mexico City.
Sabloff, Jeremy A.
  1998   Distinguished Lecture in Archaeology: Communication and the Future of
          American Archaeology. *American Anthropologist* 100(4):869–75.
Sabloff, Jeremy A., and Gair Tourtellot
  1991   *The Ancient Maya City of Sayil: The Mapping of a Puuc Region Center.* Middle
          American Research Institute Publication 60. Tulane University, New Orleans.
Said, Edward
  1979   *Orientalism.* Random House, NY.
Sanders, William T.
  1960   Prehistoric Ceramics and Settlement Patterns in Quintana Roo, Mexico. *Con-
          tributions to American Anthropology and History* 12(60). Publication 606. Carne-
          gie Institution of Washington, Washington, DC.
Sanders, William T., and Barbara Price
  1968   *Mesoamerica: the Evolution of a Civilization.* Random House, NY.

Schele, Linda, and David Freidel
  1990   *A Forest of Kings: The Untold Story of the Ancient Maya.* William Morrow, NY.
Schele, Linda, Nikolai Grube, and Erik Boot
  1998   Some Suggestions on the K'atun Prophecies in the Books of the Chilam Balam
         in Light of Classic-Period History. In *Memorias del Trecer Congresso Interna-
         cional de Mayistas* 1995:399–432. Universidad Nacional Autonoma de México,
         Mexico City.
Schele, Linda, and Peter Mathews
  1998   *The Code of Kings: The Language of Seven Sacred Maya Temples and Tombs.* Scrib-
         ner, NY.
Schele, Linda, and Mary E. Miller
  1986   *The Blood of Kings: Dynasty and Ritual in Maya Art.* Kimbell Art Museum, Fort
         Worth, TX.
Schmidt, Peter J.
  1999   Chichén Itzá resultados y proyectos nuevos (1992–1999). *Arqueología Mexi-
         cana* 7(37):32–39.
  2000   Nuevos datas sobre la arqueología e iconografía de Chichén Itzá. *Investigadores
         de la Cultura Maya* 8(1):38–48.
Schufeldt, P. W.
  1950   Reminiscences of a *Chiclero.* In *Morleyana.* School of American Research and
         Museum of New Mexico, Santa Fe.
Seler, Eduard
  1898   Quetzalcoatl-Kukulkan in Yucatán. *Zeit. für Ethnologie* 30:377–410.
Sharer, Robert J.
  1994   *The Ancient Maya.* 5th ed. Stanford University Press, Palo Alto, CA.
  2003   Founding Events and Teotihuacan Connections at Copan, Honduras. In *The
         Maya and Teotihuacan: Reinterpreting Early Classic Interaction,* ed. G. E. Braswell,
         pp. 143–65. University of Texas Press, Austin.
Shaw, Justine M.
  1998   *The Community Settlement Patterns and Residential Architecture of Yaxuná from
         AD 600–1400.* Ph.D. dissertation, Department of Anthropology, Southern
         Methodist University, Dallas.
  2001   Maya Sacbeob: Form and Function. *Ancient Mesoamerica* 12:261–72.
Shaw, Justine M., and Dave Johnstone
  1996   Core and Periphery Mapping during the 1995 Season. In *The Selz Foundation
         Yaxuná Project: Report of the 1995 Field Season,* ed. C. Suhler, pp. 41–44. Depart-
         ment of Anthropology, Southern Methodist University, Dallas.
  2001   The Late Classic at Yaxuná, Yucatán, Mexico. *Mexicon* 23(1):10–14.
Shaw, Justine M., Dave Johnstone, and Ruth Krochock
  2000   *Final Report of the 2000 Yo'okop Field Season: Initial Mapping and Surface Collec-
         tions.* College of the Redwoods, Eureka, CA.
Sheets, Payson
  1992   *The Cerén Site.* University of Colorado, Boulder.
Sidrys, Raymond
  1977   Mass-Distance Measures for the Maya Obsidian Trade. In *Exchange Systems in
         Prehistory,* ed. T. K. Earle and J. E. Ericson, pp. 91–107. Academic Press, NY.
  1978   Megalithic Architecture and Sculpture of the Maya Area. In *Papers on the Econ-
         omy and Architecture of the Ancient Maya,* ed. Raymond Sidrys, pp. 155–83. Mon-
         ograph no. 7. Institute of Archaeology, University of California, Los Angeles.

Siemans, A. H., and Dennis E. Puleston
1972 Ridged Fields and Associated Features in Southern Campeche: New Perspectives on the Lowland Maya. *American Antiquity* 37:228–39.

Sierra Sosa, Thelma N.
1994 Contribución al estudio de los asentamientos de San Gervasio, Isla De Cozumel. *Colección Científica*, Serie Arqueología 279. Instituto Nacional de Antropología e Historia, Mexico City.
1999 Xcambó: Codiciado enclave económico del Clásico Maya. *Arqueología Mexicana* 7(37):40–47.
2001 Xcambó. *Mexicon* 23(2):27.

Silva Rhoads, Carlos, and Concepcion Maria del Carmen Hernandez
1991 *Estudios de patron de asentamiento en Playa del Carmen, Quintana Roo*. Colección Cientifica, Serie Arqueología no. 231. Instituto Nacional de Antropología e Historia, Mexico City.

Silverstein, Jay
2001 Aztec Imperialism at Oztuma, Guerrero: Aztec-Chontal Relations during the Late Postclassic and Early Colonial Periods. *Ancient Mesoamerica* 12:31–48.

Sivarajasingnam, S., L. T. Alexander, J. G. Cady, and M. G. Cline
1962 Laterite. *Advances in Agronomy* 14:1–60.

Smith, J. Gregory
2000 *The Chichén Itzá–Ek Balam Transect Project: An Intersite Perspective on the Political Organization of the Ancient Maya*. Ph.D. dissertation, Department of Anthropology, University of Pittsburgh. UMI, Ann Arbor, MI.

Smith, J. Gregory, William M. Ringle, and Tara M. Bond
1998 *Recent Research at Ichmul de Morley, Yucatán, Mexico and Its Implications for Northern Lowland Maya Archaeology*. Paper presented at the 63d Annual Meeting of the Society for American Archaeology, Seattle.

Smith, Robert E.
1971 *The Pottery of Mayapan*. Peabody Museum of Archaeology and Ethnology, Harvard University, Cambridge, MA.

Smith, Robert E., and James C. Gifford
1965 Pottery of the Maya Lowlands. In *Archaeology of Southern Mesoamerica*, ed. G. R. Willey. Handbook of Middle American Indians, Vol. 2. University of Texas Press, Austin.

Smyth, Michael P.
1998 Before the Florescence: Chronological Reconstructions at Chac II, Yucatán, Mexico. *Ancient Mesoamerica* 9(1):137–50.
1999 *A New Study of the Gruta de Chac, Yucatán, Mexico*. Foundation for the Advancement of Mesoamerican Studies, Crystal River, FL.
2002 *Early Puuc Urbanism at Chac II, Yucatán, Mexico: Phase II: A Report on the 2002 Field Season*. Submitted to the National Geographic Society, Washington, DC.

Smyth, Michael P., and Christopher D. Dore
1992 Large Site Archaeological Methods at Sayil, Yucatán, Mexico: Investigating Community Organization at a Prehispanic Maya Center. *Latin American Antiquity* 3:3–21.
1994 Maya Urbanism at Sayil, Yucatán. *National Geographic Research and Exploration* 10:38–55.

Smyth, Michael P., Christopher D. Dore, and Nicholas P. Dunning
    1995   Interpreting Prehistoric Settlement Patterns: Lessons From the Maya Center of Sayil, Yucatán. *Journal of Field Archaeology* 22(3):321–47.
Smyth, Michael P., José Ligorred P., David Ortegón Z., and Pat Farrell
    1998   An Early Classic Center in the Puuc Region: New Data from Chac II, Yucatán, Mexico. *Ancient Mesoamerica* 9:233–57.
Smyth, Michael P., and Daniel Rogart
    2004   A Teotihuacan Presence at Chac II, Yucatán, Mexico: Implications for Early Political Economy of the Puuc Region. *Ancient Mesoamerica* 15:14–47.
Southall, Aidan W.
    1956   *Alur Society: A Study in Processes and Types of Domination.* W. Heffers and Sons, Cambridge.
    1988   The Segmentary State in Africa and Asia. *Society for Comparative Study of Society and History* 30(1):52–82.
Standley, Paul C.
    1920–1926   Trees and Shrubs of Mexico. In *Contributions from the United States National Herbarium, Vol. 23, Parts 1–5.* Smithsonian Institution, Washington, DC.
Stanton, Travis W.
    2000   *Heterarchy, Hierarchy, and the Emergence of the Northern Lowland Maya: A Study of Complexity at Yaxuná, Yucatán, Mexico (400 BC–AD 600).* Ph.D. dissertation, Department of Anthropology, Southern Methodist University, Dallas.
Stanton, Travis W., Traci A. Ardren, and Tara M. Bond
    2000   *Chunchucmil as a Specialized Trade Center in Western Yucatán.* Paper presented at the 65th Annual Meeting of the Society for American Archaeology, Philadelphia.
Stephens, John Lloyd
    1843   *Incidents of Travel in Yucatán.* 2 vols. Harper, NY.
Stephens, John Lloyd, and Frederick Catherwood
    1963   *Incidents of Travel in Yucatán, Volume Two.* Dover Publications, NY.
Stocker, Terry, and Dan M. Healan
    1989   The East Group and Nearby Remains. In *Tula of the Toltecs*, ed. Dan M. Healan, pp. 149–62. University of Iowa Press, Iowa City.
Stone, Andrea
    1989   Disconnection, Foreign Insignia, and Political Expansion: Teotihuacan and the Warrior Stelae of Peidras Negras. In *Mesoamerica After the Decline of Teotihuacan AD 700–900*, ed. R. A. Diehl and J. C. Berlo, pp. 153–72. Dumbarton Oaks Research Library and Collection, Washington, DC.
Story, Rebecca
    1985   An Estimate of Mortality in a Precolumbian Urban Population. *American Anthropologist* 87:519–35.
    1992   *Life and Death in the Ancient City of Teotihuacan: A Modern Paleodemographic Synthesis.* University of Alabama Press, Tuscaloosa.
Stuart, David
    2000   The Arrival of Strangers: Teotihuacan and Tollan in Classic Maya History. In *Mesoamerica's Classic Heritage: From Teotihuacan to the Aztecs*, ed. D. Carrasco, L. Jones, and S. Sessions, pp. 465–513. University Press of Colorado, Boulder.
Stuart, David, Nikolai Grube, and Linda Schele
    1989   *A Substitution Set for the "Ma Cuch/Batab" Title.* Copán Note 58. Copán Acropo-

lis Archaeological Project and Instituto Hondureño de Antropología e Historia, Copán, Honduras.

Suhler, Charles
  1996   *Excavations at the North Acropolis, Yaxuná, Yucatán, Mexico*. Ph.D. dissertation,
          Department of Anthropology, Southern Methodist University, Dallas.

Suhler, Charles, Traci A. Ardren, and David Johnstone
  1998   The Chronology of Yaxuná: Evidence from Excavation and Ceramics. *Ancient
          Mesoamerica* 9:167–80.

Suhler, Charles, and David A. Freidel
  1993   *The Selz Foundation Yaxuná Project: Final Report of the 1992 Field Season*. Depart-
          ment of Anthropology, Southern Methodist University, Dallas.
  1998   Life and Death in a Maya War Zone. *Archaeology* 51(3):28–34.

Sullivan, Paul R.
  1989   *Unfinished Conversations. Mayan and Foreigners Between Two Wars*. Alfred
          Knopf, NY.

Tambiah, Stanley J.
  1977   The Galactic Polity: The Structure of Traditional Kingdoms in Southeast Asia.
          Annals of the New York Academy of Sciences 293:69–97.

Taube, Karl A.
  1994   The Iconography of Toltec Period Chichén Itzá. In *Hidden among the Hills: The
          Archaeology of Northwestern Yucatán*, ed. Hans J. Prem, pp. 216–46. Acta Meso-
          americana 7. Verlag von Flemming, Möckmühl, Germany.
  1995   The Monumental Architecture of the Yalahau Region and the Megalithic Style
          of the Northern Maya Lowlands. In *The View From Yalahau: 1993 Archaeologi-
          cal Investigations in Northern Quintana Roo, Mexico*. Latin American Studies Pro-
          gram, Field Report Series, no. 2. University of California Press, Riverside.
  2003   Tetitla and the Maya Presence at Teotihuacan. In *The Maya and Teotihuacan:
          Reinterpreting Early Classic Interaction*, ed. Geoffrey E. Braswell, pp. 273–314.
          University of Texas Press, Austin.

Taube, Karl A., and Tomás Gallareta Negrón
  1989   *Survey and Reconnaissance in the Ruinas de San Angel Region, Quintana Roo, Mex-
          ico*. Unpublished report submitted to the National Geographic Society, Wash-
          ington, DC.

Taylor, Dicey
  1983   *Classic Maya Costume: Regional Types of Dress*. 2 vols. Ph.D. dissertation, De-
          partment of Anthropology, Yale University. University Microfilms, Ann Ar-
          bor, MI.
  1992   Painted Ladies: Costumes for Women on Tepeu Ceramics. In *The Maya Vase
          Book, Volume 3*, ed. Justin Kerr, pp. 513–25. Kerr Associates, NY.

Thompson, J. Eric S.
  1970   *Maya History and Religion*. University of Oklahoma Press, Norman.

Thompson, J. Eric S., Harry E. D. Pollock, and J. Charlot
  1932   *A Preliminary Study of the Ruins of Cobá, Quintana Roo, Mexico*. Publication 424.
          Carnegie Institution of Washington, Washington DC.

Tourtellot, Gair, III
  1988a   *Excavations at Seibal, Department of Peten, Guatemala: Peripheral Survey and Ex-
          cavation Settlement and Community Patterns*. Memoirs of the Peabody Museum
          of Archaeology and Ethnology Vol. 16. Harvard University Press, Cambridge,
          MA.

1988b Developmental Cycles of Households and Houses at Seibal. In *Household and Community in the Mesoamerican Past*, ed. Richard R. Wilk and Wendy A. Ashmore, pp. 97–120. University of New Mexico Press, Albuquerque.

Tourtellot, Gair, III, and Jeremy A. Sabloff
1989 Approaches to Household and Community Structure at Sayil. In *Households and Communities: Proceedings of the Twenty-first Annual Conference of the Archaeological Association of the University of Calgary*, ed. S. MacEachern, D. J. W. Archer, and R. D. Garvin, pp. 363–70. University of Calgary, Alberta, Canada.
1994 Community Structure at Sayil: A Case Study of Puuc Settlement. In *Hidden Among the Hills*, ed. H. J. Prem, pp. 71–90. Verlag Von Fleming, Mockmuhl, Germany.

Tozzer, Alfred M.
1957 *Chichén Itzá and Its Cenote of Sacrifice*. Memoirs of the Peabody Museum of Archaeology and Ethnology Vols. 11–12. Harvard University, Cambridge, MA.

Turner, B. L. II
1974 Prehistoric Intensive Agriculture in the Maya Lowlands. *Science* 185:118–24.

Turner, B. L., II, and Peter D. Harrison (eds.)
1983 *Pulltrouser Swamp: Ancient Maya Habitat, Agriculture, and Settlement in Northern Belize*. University of Texas Press, Austin.

Urry, J.
1990 *The Tourist Gaze*. Sage Publications, London.

Vallo, Michael
2000 Die Geschichte einer Maya-Siedlung: Forschungsergebnisse aus den Ausgrabungen von Xkipché. In *Maya Gottkönige im Regenwald*, ed. Nikolai Grube, Eva Eggebrecht, and Matthias Seidel, pp. 216–17. Könemann Verlagsgesellschaft, Cologne, Germany.

Vanden Bosch, Jon C.
1999 *Lithic Economy and Household Interdependence among the Late Classic Maya of Belize*. Ph.D. dissertation, Department of Anthropology, University of Pittsburgh, Pittsburgh.

Varela Torrecilla, Carmen
1998 La secuencia histórica de Oxkintok: Problemas cronológicos y metodológicos desde el punto de vista de la cerámica. *Revista Española de Antropología Americana*, 26:29–55.

Varela Torrecilla, Carmen, and Geoffrey Braswell
2003 Teotihuacan and Oxkintok: New Perspectives from Yucatán. In *The Maya and Teotihuacan: Reinterpreting Early Classic Interaction*, ed. G. E. Braswell, pp. 249–72. University of Texas Press, Austin.

Vargas, Ernesto, Patricia S. Santillan, and Marta Vilalta
1985 Apuntes para el analisis del patron de asentamiento de Tulum. *Estudios de Cultura Maya* 16:55–83.

Vargas de la Peña, Leticia, and Victor R. Castillo Borges
1999 Ek' Balam: Ciudad que empieza a revelar sus secretos. *Arqueología Mexicana* VII (37):24–31.
2001 La pintura mural prehispánica en Ek' Balam, Yucatán. In *La Pintura Mural Prehispánica en México II: Área Maya*, ed. Beatriz de la Fuente, 4:403–8. Universidad Nacional Autónoma de México, Instituto de Investigaciones Estéticas, Mexico City.

Vargas de la Peña, Leticia, Víctor R. Castillo Borges, and Alfonso Lacadena
   1999   Textos glíficos de Ek' Balam (Yucatán, México): Hallazgos de las temporadas
          de 1996–1998. In *Los Investigadores del Cultura Maya* 7, 1:172–87. Universidad
          Autónoma de Campeche, Campeche, Mexico.
Vargas de la Peña, Leticia, Thelma N. Sierra Sosa, and Carlos Peraza Lope
   1994   *Informe de la temporada 1994 en Ek Balam.* Centro INAH Yucatán, Instituto
          Nacional de Antropología e Historia, Mexico City.
Velázquez Morlet, Adriana, Edmundo López de la Rosa, Alejandro Pacheco Mendez,
Carlos Ruiz Ulloa, and Miguel Angel Valenzuela Tovar
   1991   Algunos comentarios sobre las caracteristicas arquitectonicas del noreste de
          Yucatán. *Cuadernos de Arquitectura Mesoamericana* 12:57–63.
Velázquez Valádez, Ricardo
   1980   Recent Discoveries in the Caves of Loltun, Yucatán, Mexico. *Mexicon* 2(4):53–
          55.
Vlcek, David T., and William L. Fash Jr.
   1986   Survey in the Outlying Areas of the Copán Region, and the Copán-Quirigua
          "Connection." In *The Southeast Maya Periphery*, ed. Patricia A. Urban and Ed-
          ward M. Schortman, pp. 102–13. University of Texas Press, Austin.
Vlcek, David T., Sylvia Garza T., and Edward Kurjack
   1978   Contemporary Farming and Ancient Maya Settlements: Some Disconcerting
          Evidence. In *Pre-Hispanic Maya Agriculture*, ed. Peter D. Harrison and B. L. Tur-
          ner II, pp. 211–23. University of New Mexico Press, Albuquerque.
Vogt, Evan Z.
   1969   *Zinacantan: A Maya Community in the Highlands of Chiapas.* Harvard Univer-
          sity Press, Cambridge, MA.
von Falkenhausen, Lorthar
   1985   Architecture. In *A Consideration of the Early Classic Period in the Maya Lowlands*,
          ed. Gordon R. Willey and Peter Mathews, pp. 111–33. Publication 10. Institute
          for Mesoamerican Studies, State University of New York at Albany.
Von Nagy, Christopher L., Mary D. Pohl, and Kevin O. Pope
   2002   *Ceramic Chronology of the La Venta Olmec Polity: The View from San Andres, Ta-
          basco.* Paper presented at the 67th Annual Meeting of the Society for Ameri-
          can Archaeology, Denver.
von Winning, Hasso
   1985   *La iconographía de Teotihuacán: Los dioses y los signos.* 2 vols. Universidad Na-
          cional Autónoma de México, Mexico City.
Walling, Stanley, M. Vogel, J. Arahill, K. Baedenkopf, R. Holmlund, S. Larkin, and J. Stahl
   1999   *Living on the Edge: Classic Maya Settlement Resource Use on the Rio Bravo Escarp-
          ment, Belize, Central America.* Paper presented at the 64th Annual Meeting for
          the Society for American Archaeology, Chicago.
Ward, W. C., A. E. Weidie, and William Back
   1985   *Geology and Hydrogeology of the Yucatán and Quaternary Geology of Northeastern
          Yucatán Peninsula.* New Orleans Geological Society, New Orleans.
Watanabe, Takeshi
   2000   *Form and Function of Metates in Chunchucmil, Yucatan, Mexico.* Master's thesis,
          Department of Anthropology, Florida State University, Tallahassee.
Webster, David
   1978   Three Walled Sites of the Northern Maya Lowlands. *Journal of Field Archaeology*
          5:375–90.

1979   *Cuca, Chacchob Dzonot Aké: Three Walled Northern Maya Centers.* Occasional
        Papers in Anthropology. Pennsylvania State University, University Park.
1985   Recent Settlement Survey in the Copán Valley, Honduras. *Journal of American
        Archaeology* 5:39–51.
1997   City-States of the Maya. In *The Archaeology of City-States: Cross-Cultural Ap-
        proaches*, ed. Deborah L. Nichols and Thomas H. Charlton, pp. 135–54. Smith-
        sonian Institution Press, Washington, DC.
Weidie, A. E.
1985   Geology of Yucatán Platform. In *Geology and Hydrogeology of the Yucatán and
        Quaternary Geology of Northeastern Yucatán Peninsula*, ed. W. C Ward, A. E. Wei-
        die, and W. Back, pp. 1–19. New Orleans Geological Society, New Orleans.
White, William B.
1988   *Geomorphology and Hydrology of Karst Terrain.* Oxford University Press, Oxford.
Wilk, Richard R.
1983   Little House in the Jungle: The Cause of Variation in House Size among Mod-
        ern Maya. *Journal of Anthropological Archaeology* 2:99–116.
1984   Households in Process: Agricultural Change and Domestic Transformation
        among the Kekchi Maya of Belize. In *Households: Comparative and Historical
        Studies of the Domestic Group*, ed. Robert McC. Netting, Richard R. Wilk, and
        Eric J. Arnould, pp. 217–44. University of California Press, Berkeley.
Wilk, Richard R., and Robert Netting
1984   Households: Changing Forms and Functions. In *Households: Comparative and
        Historical Studies of the Domestic Group*, ed. Robert Netting, Richard R. Wilk and
        Eric J. Arnould, pp. 1–28. University of California Press, Berkeley.
Wille, Sarah, Alfredo Minetti, and Christopher Andres
2000   *Sharing Knowledge and Exchanging Ideas: An Archaeology for Everyone.* Paper pre-
        sented at the 99th Annual Meeting of the American Anthropological Associa-
        tion, San Francisco.
Willey, Gordon, and W. J. Bullard
1965   Prehispanic Settlement Patterns in the Maya Lowlands. In *Archaeology of
        Southern Mesoamerica*, ed. Gordon Willey, pp. 360–77. Handbook of Middle
        American Indians, vol. 2. University of Texas Press, Austin.
Willey, Gordon R., William R. Bullard Jr., John B. Glass, and James C. Gifford
1965   *Prehistoric Maya Settlements in the Belize Valley.* Papers of the Peabody Museum
        of Archaeology and Ethnology, Vol. 54. Harvard University, Cambridge, MA.
Willey, Gordon R., T. Patrick Culbert, and R.E.W. Adams
1967   Maya Lowland Ceramics: A Report from the 1965 Guatemala City Confer-
        ence. *American Antiquity* 32(3):289–315.
Wren, Linnea H., and Peter J. Schmidt
1990   Elite Interaction during the Terminal Classic Period: New Evidence from Chi-
        chén Itzá. In *Classic Maya Political History*, ed. T. Patrick Culbert, pp. 199–225.
        Cambridge University Press, Cambridge.
Yaeger, Jason, and Greg Borgstede
2003   Professional Archaeology and the Modern Maya: A Historical Sketch. In *Con-
        tinuities and Changes in Maya Archaeology: Perspectives at the Millennium*, ed.
        Charles W. Golden and Greg Borgstede, pp. 259–85. Routledge, NY.

About the Contributors

George J. Bey III received B.A.'s in English and anthropology from the University of New Mexico and an M.A. and a Ph.D. in anthropology, with a specialization in Mesoamerican archaeology, from Tulane University. He has conducted fieldwork in Louisiana and in the Toltec region of Hidalgo, Mexico, has been co-principal investigator of the Ek Balam Project in Yucatán, Mexico since 1984, and currently is the co-principal investigator of the Kiuic Project in the same state. His recent publications include "Ceramic Change and Conflict in Maya Archaeology" in the volume *Warfare and Conflict in Ancient Mesoamerica*; "Postclassic and Terminal Classic Courts of the Northern Maya Lowlands" (with William Ringle) in the volume *Maya Courts*; and the article "The Return of Quetzalcoatl: Evidence for the Spread of a World Religion during the Epiclassic Period" in *Ancient Mesoamerica*. His research interests include the archaeological study of economics, ethnoarchaeology, regional analysis of complex societies, and pottery analysis. He is currently an associate professor and dean at Millsaps College in Jackson, Mississippi.

Tara M. Bond-Freeman received a B.A. in sociology from Millsaps College and an M.A. in anthropology from Louisiana State University and is currently working on a Ph.D. in anthropology, with a specialization in Maya archaeology, from Southern Methodist University. She worked as a ceramic analyst on the Ek Balam Project from 1992–99 and is now a member of the Chunchucmil Project in Yucatán. Her recent publications include the article "The Ceramic Chronology of Ek Balam, Yucatán, Mexico" (with others) in *Ancient Mesoamerica*, and she has presented numerous papers at professional conferences. She also has experience in museum curation and exhibition.

Rafael Cobos received a Ph.D. in anthropology from Tulane University and is a Professor at the Universidad Autónoma de Yucatán in Mérida, Mexico. His recent publications include the articles "The Late and Terminal Classic Causeway Systems of Chichén Itzá" and "Dos modelos en la explicación arqueológica"; and book chapters "Chichén Itzá: Settlement and Hegemony during the Terminal Classic Period," "Ancient Community Forms and Social Complexity at Chichén Itzá, Yucatán," and "The Late Classic Period in Southeastern Mesoamerica." He is the author of the forthcoming book *Puertos marítimos en tierras bajas Mayas* and co-author with Payson D. Sheets of the book *San Andrés y Joya de Cerén*. His research interests are centered on Classic period settlements of the Yucatán Peninsula and El Salvador and issues of social complexity, urbanism and cities, trade and exchange, and economic specialization. He is currently the director of the Sihó Archaeological Project, which focuses on the study of social complexity and the rise of civilization in western Yucatán/northern Campeche.

Kurt R. Heidelberg received a B.A. in mathematical sciences and an M.S. in computer science from Virginia Commonwealth University and an M.A. in anthropology from the University of California, Riverside. He has extensive field experience in North America and has worked with the Yalahau Regional Human Ecology Project in Quintana Roo, Mexico since 1997. His recent publications include "Hodai Sonwuinakud: Making and Using Stone Tools in the Western Papagueria" (with Christopher Doolittle and Jeffrey Altschul) and "Predictive Modeling in the Military: Similar Goals, Divergent Paths." His research includes ethnographic investigations of Maya use of domestic space, the archaeology of subaquatic caves, lithic technology, and predictive modeling and spatial statistics for archaeology. He is a Ph.D. candidate in anthropology at the University of California, Riverside.

Charles W. Houck Jr. received a B.A. from Davidson College and an M.A. and a Ph.D. in anthropology, with a specialization in Maya archaeology, from Tulane University. He has worked on the Ek Balam Project in Yucatán, Mexico since 1989. His recent publications include "The Ceramic Chronology of Ek Balam, Yucatán, Mexico" (with others) in *Ancient Mesoamerica*; "The Decline of the East: Classic to Postclassic Transition at Ek Balam, Yucatán" (with others) in the forthcoming *The Terminal Classic in the Maya Lowlands: Collapse, Transition, and Transformation*; and his Ph.D. dissertation on hinterland settlement around the site of Ek Balam. His research interests include settlement and sociopolitical dynamics in hinterland areas between major centers. He is currently teaching anthropology courses at the University of North Carolina at Charlotte.

Scott R. Hutson received a B.A. in Latin American studies from Yale University and an M.A. and a Ph.D. in anthropology from the University of California, Berkeley. Prior to fieldwork in Yucatán for the last seven years, Scott worked in Oaxaca, Peru, Belize, Guatemala, and Connecticut and conducted ethnographic fieldwork in cyberspace. His research interests include archaeological method and theory, urbanism, archaeologies of landscape and dwelling, archaeological chemistry, and the socio-politics of archaeological research. He is currently affiliated with the Archaeological Research Facility at the University of California and is an adjunct professor at the University of California at Santa Cruz.

Dave Johnstone received a B.A. and an M.A. in archaeology at Simon Fraser University and a Ph.D. at Southern Methodist University. He is a ceramic specialist and has conducted fieldwork in British Columbia, on the Yaxuná Project in Yucatán and is currently the co-director of the CRAS Project in Quintana Roo, Mexico. His recent publications include "The Ceramic Chronology of Yaxuna" (with Charles Suhler and Traci Ardren) in *Ancient Mesoamerica*; the article "The Late Classic at Yaxuná, Yucatán, Mexico" (with Justine Shaw) in the journal *Mexicon*; and his Ph.D. dissertation on the ceramics of Yaxuná. He is an adjunct instructor at Humboldt State University in northern California where he teaches courses in anthropology and archaeology.

Aline Magnoni is a graduate student of Maya archaeology at Tulane University. She received a B.A. in archaeology from the Institute of Archaeology, University College London and an M.A. in anthropology from Louisiana State University. She has conducted fieldwork in Mesoamerica and Central America for the last eleven years, participating in archaeological projects in Belize, Mexico (Yucatán, Veracruz, and Quintana Roo), and Panama. For the past ten years, she has participated in the Pakbeh Regional Economy Program at the site of Chunchucmil, Yucatán, Mexico, where she has been supervising the site-mapping operations and conducting excavations at the household level. She became assistant director in 1998. Her research interests include settlement patterns, household archaeology, economic and political organization, palaeoenvironmental and palaeoclimatic studies, and community and public archaeology.

Rubén Maldonado Cárdenas received an M.S. degree, with a specialization in archaeology, from the Antropológicas de la Escuela Nacional de Antropología e Historia (ENAH) in Mexico. While at their branch in the former Department of Prehistory, he developed projects in the Mexican Altiplano where he collaborated in Tlapacoya, in Los Reyes, La Paz, in the first excavations of the Metro of Mexico City, and in the Archaeological Rescue of the Presa de la Angostura in Chiapas. In recent years, he has acted as the director of the Dzibilchaltun Project. Some of his recent publications include "La exploración y restauración de la Subestructura 44, de Dzibilchaltún" in *Yucatán a Través de los Siglos*; "Kalom Uk'Uw, señor de Dzibilchaltún" in *La Organización Social Entre los Mayas*; "Las paletas del Infiernillo, Michoacán-Guerrero y las Hohokam del suroeste de los Estados Unidos" in *El Pasado Arqueológico de Guerrero*; and "El talud-tablero remetido de Dzibilchaltún. ¿Desarrollo regional o de influencia Teotihuacana?" in *Los Investigadores de la Cultura Maya*. At present, he is part of the Centro Regional Yucatán (CRY) of INAH, where he has worked on the ballcourt of Uxmal, the Palace of Chacbolay, and in the ruins of Aké, Ucí, and Izamal.

Daniel E. Mazeau received a B.A. from the State University of New York at Albany and an M.A. from the State University of New York at Buffalo, where he is currently a Ph.D. candidate. Trained as a lithic specialist with a specific interest in obsidian tool production and sourcing, his research interests include the analysis of past economies and the integration of micro- and macro-regional scales of production, exchange, and consumption of chipped stone tools. His dissertation is an analysis of these issues at Chunchucmil and is empirically based on the chipped stone artifact assemblages of Chunchucmil and its associated hinterland sites. He is a project director for the Cultural Resource Survey Program (CRSP) at the New York State Museum in Albany.

David Ortegón Zapata received a B.A. in archaeology, with a specialization in Maya archaeology, from the Universidad Autónoma de Yucatán. He has conducted archaeological fieldwork in the Yucatán, Mexico since 1983 and is currently the ceramist for the Chac Archaeological Project. His research interests include Maya ceramics and settlement patterns.

Alicia Re Cruz received a B.A. in history, with a specialization in American ethnology and anthropology, from the Universidad Complutense de Madrid, Spain and an M.A. and a Ph.D. in anthropology from the State University of New York at Albany. She has conducted ethnographic research among the Yucatec Maya since 1986. A product of her 1989–90 fieldwork in Chan Kom and Cancún is the ethnography *The Two Milpas of Chan Kom*, published by the SUNY Press. Since 1997, she has been involved in the ethnographic documentary, "Los Otros: Maya Migrants in Cancun," which is a collaborative project with video specialist Melinda Levin. Her research interests include displacement (migrants and refugees), transnationalism, Hispanic migrants in Texas, and the effects of tourism among native communities. She is an associate professor at the University of North Texas where she teaches courses on theory in anthropology, symbolical anthropology, migrants and refugees, Mesoamerica, Latin America, and contemporary ethnic groups.

William M. Ringle received a Ph.D. in anthropology from Tulane University. He was co-principal investigator (with George Bey) of the Ek Balam Project from 1984 to 1999 and currently is the co-principal investigator (with Tomás Gallareta Negrón and George Bey) of the Kiuic Project, both in the state of Yucatán, Mexico. He is a professor and chair of the Department of Anthropology at Davidson College.

Dominique Rissolo received a B.A. from San Diego State University and an M.A. and a Ph.D. in anthropology, with a specialization in Maya archaeology, from the University of California, Riverside. He has conducted fieldwork in Belize, El Salvador, and Yucatán, Mexico, focusing his efforts on cave archaeology, rock art, and ceramic analysis. Some of his recent publications include a chapter "Beneath the Yalahau: Emerging Patterns of Ancient Maya Ritual Cave Use from Northern Quintana Roo, Mexico" in *In the Maw* *of the Earth Monster: Studies of Mesoamerican Ritual Cave Use* (University of Texas Press); the article "The Process and Sociocultural Significance of Gopher Trapping in a Modern Yucatec Maya Community" (with Kevin Hovey) in the *Journal of Ethnobiology*; and his Ph.D. dissertation, "Ancient Maya Cave Use in the Yalahau Region" (published by the Association for Mexican Cave Studies). He is an anthropology lecturer at San Diego State University.

Justine Shaw received a B.A. from the University of Arizona and an M.A. and a Ph.D. in anthropology, with a specialization in Maya archaeology, from Southern Methodist University. She has conducted archaeological fieldwork with the University of Arizona Garbage Project, in New Mexico, with the Yaxuná Project in Yucatán, Mexico since 1994, and is currently the co-director of the CRAS Project in Quintana Roo, Mexico. Her recent publications include "Maya *Sacbeob*: Form and Function" and "Climate Change and Deforestation: Implications for the Maya Collapse" in *Ancient Mesoamerica*; and "The Late Classic at Yaxuná, Yucatán, Mexico" (with Dave Johnstone) in the journal *Mexicon*. Her research interests include road systems, settlement patterns, and the Late Classic and Terminal Classic periods. She is a professor at the College of the Redwoods in northern California, where she teaches courses in anthropology and archaeology.

J. Gregory Smith received a B.S. in anthropology and a B.A. in geography from Central Washington University and a Ph.D. in anthropology, with an emphasis in Maya archaeology, from the University of Pittsburgh. He worked in Yucatán, Mexico, primarily with the Ek Balam Project, from 1994–99 and directed his own project in the vicinity of Kulubá in 2001. His recent publications include his dissertation, *The Chichén Itzá–Ek Balam Transect Project: an Intersite Perspective on the Political Organization of the Ancient Maya* and a co-authored article with members of the Ek Balam Project in *The Terminal Classic in the Maya Lowlands: Collapse, Transition, and Transformation*. His research interests include comparative political organization, ethnographic analogy, Mesoamerica, and Bronze Age Mongolia. He is a professor of anthropology at San Antonio College.

Michael P. Smyth received a B.A. from Roosevelt University in Chicago and an M.A. and a Ph.D. in anthropology, with a specialization in Mesoamerican archaeology, from the University of New Mexico. He has conducted ethnoarchaeological and archaeological fieldwork in the Puuc region, Mexico since 1985 and is currently director of the Chac Archaeological Project. His research interests include Maya urbanism, political economy, and tropical subsistence and storage systems. He is also an archaeologist in residence and a visiting assistant professor at Rollins College in Winter Park, Florida, where he teaches courses on archaeology and cultural anthropology.

Travis W. Stanton received a B.A. from Binghamton University and an M.A. and a Ph.D. from Southern Methodist University. He has conducted fieldwork in Bolivia, Alabama, Louisiana, Mississippi, New York, and for nearly eleven years, in Yucatán, Mexico. In the past few years, he has directed archaeological research at the sites of Santa Bárbara and Xocnaceh. From 1999–2000 he taught at the State University of New York at Fredonia and Jamestown Community College. His publications include an edited volume *Ancient Mesoamerican Warfare* (with M. Katherine Brown) and several articles on Yucatec archaeology. He is an associate professor of anthropology at the Universidad de las Américas, Puebla.

About the Editors

Jennifer P. Mathews received a B.A. from San Diego State University and an M.A. and a Ph.D. in anthropology, with a specialization in Maya archaeology, from the University of California, Riverside. She has conducted archaeological fieldwork in Quintana Roo, Mexico since 1993 and is currently the co-director of the Yalahau Regional Human Ecology Project. Her recent publications include the articles "Models of Cosmic Order: The Physical Expression of Sacred Space Among the Ancient Maya" (with Jim Garber) in *Ancient Mesoamerica*, "Radiocarbon Dating of Mortar and Charcoal Inclusions in Architectural Mortar: A Case Study in the Maya Region, Quintana Roo, Mexico" in the *Journal of Field Archaeology*, and "Megalithic Architecture at the Site of Victoria, Quintana Roo" in *Mexicon*. She is also the co-editor (with Justine Shaw) of the volume *Quintana Roo Archaeology* published by the University of Arizona Press. Her research interests include ancient Maya architecture, site layout, road systems, historic archaeology, and current issues that indigenous peoples are facing with the rise of tourism and interest in archaeological resources. She is an associate professor at Trinity University in San Antonio, Texas, where she teaches courses in archaeology and physical anthropology. Mathews is also a chapter contributor with a solo-authored chapter and a co-authored chapter.

"""

Bethany A. Morrison received a B.A. from the University of California, Santa Barbara and an M.S. and a Ph.D. in Anthropology, with a specialization in Maya archaeology, from the University of California, Riverside. She has conducted fieldwork with the Belize River Archaeological Settlement Survey, as well as with the Yalahau Regional Human Ecology Project in Quintana Roo, Mexico. Her recent publications include "Initial Evidence for Use of Periphyton as an Agricultural Fertilizer by the Ancient Maya Associated with the El Edén Wetland, Northern Quintana Roo, Mexico" (with Roberto Cózatl Manzano) in the volume *The Lowland Maya Area: Three Millennia at the Human-Wildland Interface* and the article "Wetland Manipulation in the Yalahau Region of the Northern Maya Lowlands" (with others) in the *Journal of Field Archaeology*. Her research interests include settlement patterns, human ecology, community structure, ancient agriculture, the use of the scientific method in archaeology, and the development of innovative teaching methods. Morrison is also a chapter contributor with a solo-authored chapter.

Index

Abandonment: in the Late Formative, 27, 31; of rituals, 219

Acanceh (Yucatán) site, 17, 20, 149

Actun Toh (Quintana Roo) site, 102, 106, 108. *See also* Caves

Aerial photography, 48, 53, 59, 61, 63, 64, 69, 78

Agriculture: domestic crops, 44; house-lots, 193, 195; maize, 188; modern, 45, 188; swidden, 40–43, 44; wetland, 44–45, 54–55

Aguadas, 68. *See also* Cenotes

Aké (Yucatán): architecture, 31, 32, 97, 98, 99, 100, 107, 110–11, 115; in the Early Classic, 17; in the Formative, 19; occupation, 107; site, 13, 34, 97, 100, 101, 106, 109, 110, 114, 117

Albarradas (stone walls): at Chunchuc-mil, 79, 82, 83, 85, 86, 87, 88; at Cobá, 81; at Cozumel, 81; at Mayapan, 81; as property walls, 192; at Xamanha, 81; at Xcaret, 81. *See also* Stone walls

Andrews, Anthony P., 20, 22, 58, 97

Andrews, E. Wyllys, IV, 17

Andrews, E. Wyllys, V, 17, 18–19, 21–22, 23, 24

Archaic period, 22, 23; states, 25, 160

Architecture: at Actun Toh, 108; at Aké, 114; blocks, 6; ceremonial, 25; at Chac II, 117, 120–21; changes in, 149; at Chan Pich, 115; at Chichén Itzá, 162–63, 165, 167; at Chunchucmil, 33, 79, 83, 85–86, 87, 88–89, 91; Classic period, 97; at Cobá, 150; dating of, 6, 100, 101, 102–28; designs, 98; at Dzibilchaltún, 27, 30, 86, 146, 151; Early Classic, 32, 34, 101, 117, 120, 150; at Ek Balam, 65, 69, 72, 101, 163–64; elements, 107, 177, 183; at El Naranjal, 107–8, 113; excavation of, 98; features, 175, 197; fill, 89; of the

Formative period, 19, 20, 23, 26–27, 30, 36, 101; groups, 84, 124, 162; at Ichmul de Morley, 170; at Ikil, 109; at Kantunilkin, 110; at Kom Chen, 29; lack of, 5; layout, 79; at Makabil, 49, 50; Megalithic style, 31, 97, 98–99, 100–106; modification of, 107; monumental, 20, 26, 51, 96; occu-pation, 100; at Oxkintok, 32, 110–11, 147, 152; at Ox Mul, 111; and politics, 57; public, 72; Puuc style, 137, 151; remains, 67; Rio Bec style, 97, 118; at San Angel, 111–12; stratigraphy, 127; structure, 29; style, 6, 31–34, 95–101, 106–18, 121, 135–38, 142, 145–46, 149–50, 163–64; at Ucí, 113; at Victo-ria, 113–14; volume, 87; at Xcambó, 114; at Xuilub, 67; at Yaxhom, 114; at Yaxuná, 30, 33, 115, 147, 149, 151, 152. *See also* Chenes architectural style; Domestic elements, architecture

Ardren, Traci, 9, 77, 199

Atlas Arqueológico de la Republica Mexi-cana, 58, 69, 77, 156, 160

Atlatl (throwing spear), 124, 136

Bajos (natural depressions), 34, 44, 45

Balankanche caverns, 181–82. *See also* Caves

Balanza ceramic type, 104, 105, 111, 114

Ball, Joseph, 17, 22, 28, 140, 155

Ballcourts: at Chichén Itzá, 162, 163, 164, 167, 179, 180; at Chunchucmil, 79; at Ek Balam, 116, 163; at Labná, 19; Middle Formative, 19, 20, 25, 30

Belize, 21, 44, 188

Bey, George J., III, 3, 74, 116, 155, 166

Bichrome: Carolina, 26, 30, 100; ceram-ics, 18, 30; Huachinango, 27, 100; Valladolid, 27, 103

Botanical remains, 42, 43

Braswell, Jeffrey, 37, 165, 180
Brechas (pathways), 59. *See also* Transects
British Honduras. *See* Belize
Burials: at Chac II, 123, 124, 133, 134,
    135, 138; at Chunchucmil, 83, 89;
    contents, 98; lack of, 30, 98, 118; at
    Oxkintok, 111; practices, 96; tombs,
    30, 111, 147; at Xuilub, 67

Cacao (*spp. Theobroma cacao*), 43, 65,
    74–75, 88, 133. *See also* Cultivation
Calakmul (Campeche) site, 110, 137, 142,
    143, 146, 152
Campeche, 44, 148, 182
Cancún: businesses, 188, 196, 216, 219;
    city, 204, 211, 212, 217, 218; highway,
    213; INAH office in, 9, 202; out-
    migration to, 8, 211–13, 215–19; as
    resort, 199, 210; tourism in, 211, 212,
    218
Carolina: Bichrome incised, 26, 30, 100,
    103, 104, 105, 111, 112; ceramic group,
    100, 112, 113, 114
Castillo at Chichén Itzá, 162, 163, 176
Causeways, 2, 16, 66, 81, 177. *See also*
    Sacbeob
Caves: Balankanche, 181; Carwash, 32;
    Chac, 6, 139; Loltún, 19, 22, 30, 114;
    Mot Mot, 108; in the Puuc region, 22;
    in the Yalahau region, 20
Cehpech: ceramics, 65, 114, 125, 136,
    169, 170; ceramic sphere, 18, 35, 151,
    164–65
Celestún (Yucatán), 88
Cenotes (karstic sinkholes): *aguadas*, 68;
    associated with sites, 51; Azul, 48,
    50, 52, 53, 54, 55; at Carmelita, 50;
    at Chan Kom, 213, 219; at Ek Balam,
    62–63, 69, 70, 73; as feature, 61, 62–
    65, 73, 75; free-zones, 61, 69, 73; at
    Makabil, 60; at Maní, 19; of Sacrifice,
    61; survey of, 50; at Tikintzec, 73; at
    X-Huyub, 73; at Xk'ek'en, 63; Xtoloc,
    Zona de, 61; at Xuilub, 73; Yohd-
    zadz, 73. *See also* Dolines; *Dzadzob*;
    Nauahuelas; *Rejolladas*
Ceramics: at Actun Toh, 108; at Aké, 107;
    Bichrome, 30; Cehpech, 18, 35, 65,
    125, 136, 151, 164, 169; at Chac II,
    125, 134, 135, 138, 139; at Chichén
    Itzá, 7, 164, 175–76, 177, 181, 182;
    chronology, 97, 98, 101; at Chun-
    chucmil, 77, 83, 89; at Ciudad Mario
    Acona, 109; at Cobá, 26, 33; dating,
    6, 65, 100, 101, 102–5, 118, 125, 136;
    at Dzonot Aké, 109; Early Classic,
    18, 33, 35, 83, 111; Early Formative
    period, 23; at Ek Balam, 26, 65, 101,
    116; Ek complex, 3; at El Naranjal,
    107; Formative period, 19, 26, 30, 35;
    at Ichmul de Morley, 159, 160, 164,
    165, 169; at Ikil, 110; at Izamal, 107; at
    Kantunilkin, 110; at Komchen, 19, 24;
    Late Classic, 149; Late Formative, 29,
    30, 108; at Makabil, 48; Middle Clas-
    sic, 111; Middle Formative, 20, 21, 22,
    23, 24, 67; Nabanche complex, 24;
    at Oxkintok, 111; at Ox Mul, 111; of
    the Petén, 21, 33; polychrome, 18, 33;
    production, 145; at Rio Azul, 145; at
    San Angel, 111–12; at San Cosmé, 112;
    Sotuta, 160, 164, 165, 169; spheres, 6,
    97, 98, 143–44; at Teotihuacan, 140;
    Terminal Classic, 151, 164; as trade
    items, 96, 145; at Tres Lagunas, 112;
    at Tula, 177, 180, 181, 182; at Ucí, 113;
    at Victoria, 113–14; at Xcambo, 114;
    at Yaxuná, 115, 145, 147, 149, 150,
    151–52
Cetelac, 104, 105, 111, 113, 114, 152
Chac: Cave, 6; effigy, 203; masks, 166;
    Slab, 127–34
Chacmool (sculpture), 166, 176, 180
Chac II (Yucatán): architecture, 32, 100;
    research at, 13, 37, 120–24; site, 6, 34,
    101, 102, 106, 116, 117
Chan Kom (Yucatán): out-migration,
    212, 216–18; research at, 211, 212,
    214–15; and tourism, 8, 212, 213,
    218–20; town, 8, 212, 213. *See also*
    Chichén Itzá
Chenes architectural style, 97, 118, 163.
    See also Megalithic style; Rio Bec
Chichen ceramics, 182
Chichén Itzá (Yucatán): architecture, 77,
    173, 175–76, 178–80, 183; ceramics,
    65, 153, 164–65, 177, 180–82, 183;
    compared with Ek Balam, 7, 65, 74,

116, 156, 160, 162–72; compared with
Tula, 7, 173–83; in the Early Postclassic, 16, 182; emblem glyph, 143;
obsidian, 165–66, 180; Old Chichén,
162, 176; politics, 143, 146, 147, 155;
as related to Chan Kom, 8, 213, 215;
as related to Ichmul de Morley, 160,
162–72; as related to Yaxuná, 143,
152, 153; research at, 13, 58, 142, 214,
215; Sacred Cenote, 61; site, 8, 109,
153, 173, 213, 215; in the Terminal
Classic, 15, 153, 173, 182; tourism
at, 200, 212. *See also* Chan Kom; Ek
Balam; Ichmul de Morley; Tula
Chikinchel: project, 15; region, 17, 26,
34, 58, 61, 75, 116
Chiquila ceramic, 100, 103, 110
Chocolate. *See* Cacao
Chultun (water storage feature), 122, 124,
135. See also *Aguadas; Cenotes*
Chumbek (Yucatán) site, 70, 71, 72, 75
Chunchucmil (Yucatán): *albarrada*, 79–
81, 87, 88; architecture, 77, 78, 79–81,
86; ceramics, 77; in the Early Classic,
32, 33, 37, 77–78; houselots, 81–92;
in the Late Classic, 78; in the Late
Formative, 28; in the Middle Formative, 3, 77; obsidian, 89–90; research
at, 5, 13, 34, 37, 77–92; site, 28, 34,
77; in the Terminal Classic, 78; trade,
33, 78–79. *See also* Pakbeh Regional
Economy Program
Ciudad Mario Acona (Quintana Roo)
site, 106, 108–9
Classic period: *albarradas*, 81; ceramics,
26, 33, 149, 164; at Chichén Itzá, 174,
175, 177; Chunchucmil, 5, 77, 78, 88,
92; cultural sphere, 97; lack of texts,
17, 18, 35; politics, 6, 142; research,
16; sites, 26, 98
Cobá (Quintana Roo): *albarradas*, 49, 81;
in the Classic period, 35; in the Early
Classic, 37; hieroglyphic texts, 143;
houselots, 85; interaction with, 74; in
the Late Classic, 144, 149, 150; Megalithic architecture, 31, 102, 106, 115;
in the Middle Formative, 19; polity,
149, 153; research at, 13; sacbe, 115,
146, 150, 151; site, 116, 213; in the

Terminal Classic, 146; and Yaxuná,
146, 149, 150
Cobos, Rafael, 5, 7, 177, 180
Copán (Honduras) site, 78, 119, 136
Corbelled features: aprons, 99, 111, 113;
vaults, 97, 100, 101, 102–4, 105, 107,
109, 110, 118, 150
Corn, 41, 42, 43, 46, 84, 88, 216, 217. *See
also* Cultivation; Maize; *Milpas*
Costa Maya Project, 15, 19, 20, 21, 27, 28
Cuevas (caves), 62. *See also* Caves
Cultivation, 41, 42, 44, 46, 51, 74, 75,
217. *See also* Cacao; Corn; Maize
Cupul Survey Project, 15, 58, 69, 160

Dahlin, Bruce, 77
Dolines, 61, 62–63, 64. *See also Cenotes*
Domestic elements: activities, 190; animals, 193, 194; architecture, 19, 79,
123, 163, 164, 170; assemblages, 120;
context, 165; crops, 44; landscape,
189; organization, 29; ritual, 30;
space, 84, 188, 192, 194, 196, 197. *See
also* Agriculture
Dunning, Nicholas, 26, 28, 31, 58, 116,
156
Dzadzob (sink holes), 62, 63, 64, 65, 69,
73, 74, 75. *See also Aguadas; Dolines;
Nauahuelas; Rejolladas*
Dzibilchaltún (Yucatán): architecture,
27, 30, 146, 151; collapse, 17, 56; emblem glyph, 143; in the Late Formative, 27, 28; occupation, 16; research,
17, 56; site, 13, 27, 35, 86
Dzilam ceramics, 30, 104, 105, 113, 114,
138
Dzonot. See Cenotes
Dzonot Aké (Yucatán) site, 102, 106, 108,
109, 153

Early Classic period, 15, 17, 18, 95, 98,
183; Actun Toh, 108; Aké, 34, 107;
architecture, 32; ceramics, 18, 30,
33, 35, 83, 101, 107, 110, 113–16, 136;
Chac II, 37, 117, 120, 124, 130, 134–
35, 140–41; Chikinchel region, 34,
115–16; Chunchucmil, 33, 34, 37,
77–78, 83, 88; Ciudad Mario Acona,
108, 109; civilization, 15; Cobá, 37;

Early Classic period (*continued*)
corbelled vault, 100, 101, 107; Dzonot
Aké, 108, 109; Ek Balam, 34, 116, 117;
El Naranjal, 34, 107; glyphic texts,
35, 36, 143; Ikil, 108, 109–10; interaction, 5, 31, 33, 95, 98, 118, 119–41;
Izamal, 34, 107; Kantunilkin, 34, 108,
110; Megalithic architecture, 6, 31–
32, 34, 100–118; obsidian, 90, 139,
180; occupation, 17, 35; Oxkintok,
32–33, 108, 110–11, 140; Ox Mul,
108; Puuc region, 6, 139; radiocarbon dates, 101, 107; research, 31–37;
San Angel, 108, 111–12; San Cosmé,
108; sites, 16; Teotihuacan, 130, 137,
139, 173, 174; tomb, 111; trade, 33,
96, 97; transition, 31; Tres Lagunas,
108; triadic groups, 100, 101; Uaxactun, 129; Ucí, 108, 113; Victoria,
108, 112–13, 114; Xcambo, 33, 34, 37,
108, 114; X-Huyub, 72; Yaxhom, 114,
115; Yaxuná, 33, 108, 115, 147, 150;
Yo'okop, 143
Early Formative period, 21, 23
Ejidos (common lands), 188, 201, 202,
204, 206, 207, 208, 214
Ek Balam (Yucatán): architecture, 32,
66, 70, 72, 73, 161, 162–64; *cenotes*,
63, 64, 73; ceramics, 3, 24, 26, 30,
65, 67, 164–65, 169, 170; compared
with Ek Balam, 7, 65, 74, 116, 156,
160, 162–72; in the Early Classic, 34;
Ek Balam Project, 15, 156, 164, 166,
171; emblem glyph, 143; hinterland
survey, 65–70, 73–76, 156; iconography, 165–68, 170–71; in the Late
Classic, 117; Megalithic architecture,
100, 101, 106, 115, 116; in the Middle
Formative, 19; obsidian, 165–66, 170;
polity, 74, 76, 155; region, 17, 26, 27,
28, 35, 62, 64; *rejollada*, 64; as related
to Ichmul de Morley, 15, 156, 160,
162–72; research at, 15, 35, 58, 59, 61,
65; *sacbe*, 66; site, 7, 13, 26, 27, 28, 35,
56, 62, 64, 66, 116; in the Terminal
Classic, 153. *See also* Chichén Itzá;
Ichmul de Morley
El Chayal obsidian, 139, 165, 166
El Edén: ecological reserve, 46, 50, 52,

53, 54, 61; wetlands, 45, 46–48,
50–52, 53, 54, 55
El Naranjal (Quintana Roo): excavations,
118; Megalithic architecture, 99, 100,
106, 107, 109, 110, 112, 113; *sacbe*, 112;
site, 34, 99, 101–4, 106–7, 111, 117,
202; wetlands, 45. *See also* Naranjal
(town)
Emblem glyphs, 142, 143, 170, 171
Epiclassic period, 182, 183
Esteban Amador, Fabio, 111, 113

Fedick, Scott L., 51, 110, 112, 156, 201
Fine Orange ceramics, 128, 130, 182
Formative period: ceramics, 19, 35, 181;
communities, 19; Dzibilchaltún,
31; El Naranjal, 191; interaction, 97;
Komchen, 31; occupation, 15, 19;
Olmec, 173; period, 3, 15, 16, 18;
research, 30; sites, 20, 26, 29, 197
Freidel, David A., 146, 150

Gallareta Negrón, Tomás, 20, 111, 112,
160
Garza Tarazona de González, Silvia, 155,
156. *See also* Yucatán Atlas Project
Glover, Jeffrey, 111, 112, 113
Gods: K, 166; Y, 130
Great Ballcourt of Chichén Itzá, 162,
167, 179, 180
Gulf Coast: ceramics, 24, 136, 182;
region, 25, 78, 137, 139, 141, 175;
tradition, 179, 181

Harrison, Peter, 44
Headdresses, 127, 129, 133, 167
Heidelberg, Kurt, 7, 55, 201
Hinterland: at Ek Balam, 5, 65–76, 156;
populations, 5; in Puuc region, 26;
settlement, 49, 56–58; studies, 58–76;
survey, 5, 42, 56–65
Holbox Fracture Zone, 41, 44
Household Garden—Residence Association (HGRA) model, 7, 187, 194, 197
Houselots: for animal husbandry, 193,
194, 195; at Chunchucmil, 5, 77, 79–
83, 85–89, 91; contact period, 196; as
garden, 194, 196; at Makabil, 49; at
Naranjal, 188–90, 192, 196; periphy-

ton used in, 55; structure, 195, 197; in Veracruz, 187

Huachinango: ceramic, 27, 100, 103, 104, 105, 111, 113, 114, 116; Incised-Bichrome, 30, 100

Huntichmul (Yucatán) site, 100, 103, 106, 115

Ichmul de Morley (Yucatán): architecture, 162–64, 169, 170; ceramics, 159, 160, 164–65, 170; and Ek Balam, 160–72; iconography, 166–68, 170; obsidian, 165–66, 170; research at, 156, 160; site, 7, 155, 157, 159, 161. *See also* Chichén Itzá; Ek Balam

Ikil (Yucatán) site, 103, 106, 108, 109

Incensario (incense burner), 181, 182. *See also* Censer

Instituto Nacional de Antropología e Historia (INAH): archaeologists, 203, 204; Cancún office, 202; curation, 202; permits, 202; projects, 15, 107, 116, 118, 162, 163, 165, 167; regulations, 98, 207

Interaction spheres: in the Late Preclassic and Early Classic, 6, 33, 34, 35, 95, 118; model, 6, 95–96, 98, 118, 174; research, 97–98

Isla Cerritos (Yucatán): 20, 169, 181

Itzá Maya, 152, 169, 170, 175

Izamal (Yucatán): in the Early Classic, 17, 34; masks, 114; Megalithic architecture, 31, 32, 33, 97, 98, 103, 106–7, 114; in the Middle Formative, 19; project, 13; site, 31, 32, 33, 97, 98, 103, 106–7, 114

Jade, 24, 25

Johnstone, Dave, 6, 17, 97, 144

Kaminaljuyu (Guatemala) site, 136

Kantunilkin (Quintana Roo) site, 34, 100, 103, 106, 108, 110, 111, 112

Kepecs, Susan, 34, 75, 170

Kinich Naranja ceramic, 134, 136

Kiuic (Yucatán): ceramics, 24, 25; in the Late Formative, 27; in the Middle Formative, 19, 20. *See also* Labná-Kiuic Regional Archaeological Project

Komchen (Yucatán): architecture, 27; ceramics, 19, 21, 24, 26; collapse, 17, 31; in the Late Formative, 17, 18, 19, 27, 28; in the Middle Formative, 18, 19, 23; project, 13, 18; *sacbe*, 29

Kukulkan (feathered serpent), 175, 176, 180, 220. *See also* Quetzalcoatl; Serpent

Kurjack, Edward, 58, 86, 155, 156, 178, 179. *See also* Yucatán Atlas Project

Labná (Yucatán) site, 13, 19, 24, 26, 58, 129

Labná-Kiuic Regional Archaeological Project, 26. *See also* Kiuic; Labná

Lankin Impressed ceramic, 149

Late Classic period: architecture at Chac II, 123–24, 135; architecture at Yaxuná, 150; Megalithic architecture, 32, 101, 103, 116–17, 121; ceramics, 32, 35, 130, 144, 149; at Chac II, 117, 120–24, 129–30, 135, 140; at Chunchucmil, 78, 83; at Cobá, 144, 150; at Dzibilchaltún, 146; at Ek Balam, 65–69, 101–3, 116; interaction spheres, 97, 140, 146; role of the northern lowlands prior to, 16–35; at Yaxuná, 149, 150, 153

Late Formative period: at Actun Toh, 102, 108; at Aké, 107; ceramics, 100, 110, 112–14; at Dzonot Aké, 102, 108, 109; around Ek Balam, 72; at Huntichmul, 103; interaction spheres, 95–118; at Izamal, 103, 107; at Kantunilkin, 103, 108, 110; at Makabil, 48; Megalithic architecture, 95–118; at Naranjal, 103, 107; at Ox Mul, 104, 108; at Oxkintok, 104, 108, 111; radiocarbon dating, 101; role of the northern lowlands, 15–36; at San Angel, 104, 108, 112; at San Cosmé, 104, 108, 112; at Sihó, 104; at Tres Lagunas, 104, 108, 113; at Ucí, 105, 108, 113; at Victoria, 105, 108, 114; wetland management, 48; in the Yalahau region, 48; at Yaxom, 105, 108; at Yaxuná, 115. *See also* Late Preclassic period

Late Postclassic period: at Actun Toh,

Late Postclassic period (*continued*)
108; at Chac II, 140; at El Naranjal,
107; interaction with Teotihuacan,
140; Megalithic architecture, 107,
108, 111–13, 116, 117; modifications
to architecture or monuments, 107;
at Ox Mul, 111; role of the northern
lowlands, 16; at San Angel, 111, 112;
at Tres Lagunas, 113
Late Preclassic period: 5, 6, 111. *See also*
Late Formative period
Leveling platforms, 126, 127, 132, 134
Leveling/Stela Platform, 131, 132, 135
Lincoln, Charles, 107, 177, 180
Lintel Building at Chac II, 124–35
Loltún Cave, 19, 22, 30, 114
Looting, 109, 110, 112, 113, 114

Maize, 192, 193, 206; supplements to,
65, 188. *See also* Corn; *Milpas*
Makabil (Quintana Roo) site, 4, 48–54
Maldonado Cárdenas, Rubén, ix, 5, 6,
32, 107, 113, 178
Matacapan (Veracruz) site, 138
Mathews, Jennifer P., 5, 8, 31, 33, 34, 113
Mayapán (Yucatán) site, 13, 16, 49, 56,
81
Megalithic style, ix, 6, 31–34, 95–118,
121–25, 127, 135, 147. *See also* Architecture; Puuc; Rio Bec
Mérida: artifacts housed in, 167; nearby
cenotes, 19, 61; nearby sites, 106, 107
Mesoamerica: interaction within, 119–
40, 173–86; trade by Maya with, 24,
34, 78
Metates (grinding stones), 88, 91, 108
Midden deposits, 89, 125, 150, 165
Middle Classic period: ceramics, 101, 111,
125, 134; at Chac II, 117, 123, 125,
134, 138; at Ek Balam, 101, 116; interaction with Teotihuacan, 137–39;
Megalithic architecture, 101, 104, 111,
117, 166; obsidian, 139; at Oxkintok,
104, 111; at Yaxuná, 147–49
Middle Formative period: ceramics,
21–24, 35, 65–67, 113–14; at Chunchucmil, 77; at Ek Balam, 65, 67;
long-distance trade, 24–25; Megalithic architecture, 113, 115; role

of the northern lowlands, 18–36;
at Tres Lagunas, 113; at Victoria,
113; at Yaxuná, 115. *See also* Middle
Preclassic period
Middle Preclassic period, 111. *See also*
Middle Formative period
Milpas (corn fields): agriculture, 42, 43,
46, 50, 51, 54, 66; city conceptualized
as, 212; modern Maya relationship to,
8, 216–20; preparation of seedlings
for, 190; settlement surveys within,
50, 59, 67; use of periphyton within,
51. *See also* Agriculture; Corn; Maize
Modified Florescent, 163, 170
Monumental architecture: at Chac II,
121, 124, 136, 138; at Chunchucmil,
5, 77; at Ek Balam, 162; Late Forma-
tive, 36; Middle Formative, 20, 23; at
Yaxuná, 27, 30, 150. *See also* Chenes
architectural style; Megalithic; Puuc;
Rio Bec
Monuments: lack of, in the northern
lowlands, 35, 98, 118; and northern
lowland politics, 143; at Oxkintok,
32, 110; at Yaxuná, 149, 151. *See also*
Stelae
Moon Lord, 130
Morley, Sylvanus G., 142, 159
Morrison, Bethany A., 4, 51
Motul ceramic complex, 113, 125, 134,
136, 144
Muna Slate ceramic group, 107, 114, 116,
135, 165
Mural painting, 111, 129, 166
Muuch group, 80, 82, 85, 88–90

Nabanche ceramic complex, 22–24
Naranjal (Quintana Roo town), 7, 107,
113, 188–97, 198–209. *See also* El
Naranjal site
Nauahuelas (*cenotes* with little water), 50,
51. *See also Cenotes*
New Chichén at Chichén Itzá (Yucatán),
162, 176
Nolo Red, 105, 113
North Acropolis at Yaxuná (Yucatán), 27,
115, 150, 151
Northern Mamon complex, 22
Nucuchtunich (Yucatán) site, 105, 114

Obsidian, 136, 139, 165–66, 170, 176, 180

Old Chichén at Chichén Itzá (Yucatán), 162, 176

Oxkintok (Yucatán) site, 13, 34, 35, 37, 137, 140–41; influence at Chac II, 135; influence at Yaxuná, 147, 149, 152; Megalithic architecture at, 31, 32–33, 104, 106, 108, 110; relation with Chunchucmil, 78

Ox Mul (Quintana Roo) site, 32, 104, 106, 108, 111

Pachuca obsidian, 136, 139, 165–66, 170, 176, 180

Pakbeh Regional Economy Program (PREP), 77. See also Chunchucmil

Palenque (Chiapas) site, 130, 171

Paleo-Indian period, 22–23

Periphyton, 4–5, 46–55

Petén, 21, 33–34, 44–45, 114, 138, 147, 149, 151

Pilgrimage, 21, 108, 139, 140, 215

Plano-relief, 128, 130, 134

Platform groups: at Tres Lagunas, 104; at Chac II, 123, 124, 136

Politics: alliances, 25, 133, 145, 153, 158, 171; boundaries, 78, 156, 158–59, 171; change, 146–52; changing viewpoints, 15, 19, 28, 32, 36; at Chunchucmil, 79, 91–92; class, 75, 95, 143; Classic, 95–118, 119, 141, 143, 149; collapse, 17, 42, 153, 180; conjunctive approach, 146, 153; control at Ichmul de Morley, 160–72; control by Calakmul, 142–43; control by Chichén Itzá, 160–72; control by Ek Balam, 160–72; at Ek Balam, 69–76, 143; factions, 200, 207, 212–20; geographic units, 142, 155–56; hegemonic control strategy, 158; and hostility, 151–52; influence of Cobá, 149–53; influence of Toltecs, 173–83; interaction, 95–118, 119–40, 146, 147–52, 173–86; interaction sphere model, 95–97, 118; Late Formative, 30, 95–118; Middle Formative, 25; reflected in architecture, 95–118, 146; reflected in sacbeob, 146; reflected in settlement patterns, 56–60, 156;

role of Teotihuacan at Chac II, 137–41; Terminal Classic, 146, 151–53; territorial control strategy, 158; titles, 142, 143; at Yaxuná, 115, 142–54

Polvero Black ceramics, 103, 105, 113

Population: decline, 31, 152; dynamics, 34, 152; estimates, 33, 72, 78; fluctuations, 31, 78; growth, 26, 28; modern, 197, 200; movements, 16, 22, 152, 177; organization, 59, 66, 76; pressure, 29, 41–42; settlement, 5, 59, 77; studies, 57; support of, 41–46, 56, 77, 78

Pueblos (towns), 8, 79, 154, 214

Puuc: architectural style, 31–32, 97, 110, 115–16, 121–23, 125, 134, 151, 163; ceramic group, 35, 164; Classic, 97; Early Classic, 31–34, 35; Late Formative, 25–28; Middle Formative, 19–20; Paleo-Indian, 22; region, 6, 15–17, 36, 58, 120, 137; segmentary state, 155; sites, 6, 20, 117; and Teotihuacan, 137, 139–41; Terminal Classic, 35; trade, 24

Pyburn, Anne K, 199, 208

Quetzalcoatl (deity), 175–80. See also Kukulkan; Serpents

Quintana Roo; ethnoarchaeology, 187–97; interaction sphere, 97; Megalithic architecture, 32, 99, 106–15; migration to, 21; Naranjal, 187–97, 198–209; wetland studies, 44–45; Yalahau region, 20, 41–45, 187–97

Radiocarbon dates, 27, 65, 101–2, 107, 121, 122, 124, 127

Rancho Ahkat, 70–73

Rancho Grande, 70–72

Rancho Xeb, 70–75

Rathje, William L., 89

Rejolladas (dry sinkholes), 62–67, 75. See also Cenotes

Ringle, William, ix, 5, 6, 25, 27–30, 74, 160, 166, 170, 179

Rio Bec style, 97, 118. See also Architecture

Rissolo, Dominique, x, 7–8, 55, 108, 201

Roads: as indicators of interaction, 115,

Roads (*continued*)
    146; as indicators of polities, ix, 6, 29,
    97, 153; as indicators of social com-
    plexity, 20; modern, 196, 212; within
    sites, 79, 150. *See also Albarradas*;
    Causeways; *Sacbeob*
Robles Castellanos, Fernando, 20, 25, 97
Rock alignments, 46–47

Saban: ceramic group, 103, 104, 111, 112,
    113; type Saban Coarse, 104, 112. *See
    also* Tancah ceramic group
Sacbeob (raised roads): at Chichén Itzá,
    162; at Chunchucmil, 79, 80; at
    Cobá, 115; at Ek Balam, 66; as indi-
    cator of complexity, 20; as indicator
    of interaction, 6, 29, 112, 146–47; as
    indicator of polities, 6, 107, 146–47;
    linking El Naranjal and San Cosmé,
    112; linking Izamal and other sites,
    107; linking Yaxuná and Cobá, 115,
    150, 153; at Yaxuná, 150. *See also*
    Causeways; Roads
Sacta Group, 123, 124, 135, 136
Salt, 28, 33, 78–79, 88
San Angel (Quintana Roo), 104, 106, 108,
    111
San Carlos (Yucatán) site, 70–73
San Cosmé (Quintana Roo) site, 104,
    106, 108, 112
Sanders, William T., 110
Sascab (soft limestone additive), 84–85,
    88
Sascabera (pockets of *sascab*), 85, 88,
    91–92, 134
Savannas, 52, 78–79. *See also* Wetlands
Saxche Orange polychrome ceramics,
    149, 154
Sayil (Yucatán) site, 13, 86–87, 117, 120
Schele, Linda, 130, 143
Serpents, 121, 130–32, 163, 166, 176–80,
    218, 220. *See also* Quetzalcoatl
Settlements: at Chac II, 120–21; at Chun-
    chucmil, 77–92; Early Classic, 6; at
    Ek Balam, 69–76, 156; between Ek
    Balam and Chichén Itzá, 156; at El
    Edén, 46–51; Formative, 19, 26, 28,
    36; hinterland, ix, 5, 46–51, 56–76; at
    Ichmul de Morley, 160; as indicator

of political structure, 169–71, 177;
    modern, 4, 214; Puuc, 6, 120; in re-
    lation to water, 46–51, 60–76; survey
    of, 42, 56–59, 66–67, 142, 147; urban,
    ix, 5, 77–92; at Yaxuná, 147–50
Shangurro Red/Orange ceramics, 103,
    105, 113
Shaw, Justine A., ix, 5, 6, 17, 97
Shells, 24, 53, 78, 90, 124, 130, 132
Shrines, 75, 80, 81, 89, 189
Sierra Red ceramics, 100, 102, 103, 104,
    105, 111, 112, 113, 114, 180
Sihó, 100, 104, 115
Silho Fine Orange ceramics, 164, 176,
    180–82
Site 38 (Quintana Roo), 100, 104, 106,
    115
Smith, J. Gregory, ix, 5, 6, 156
Smith, Robert, 164, 177, 182
Smyth, Michael, ix, 5, 6, 32, 117
Snakes. *See* Serpents
Soil: augmentation, 5, 46, 53, 54, 86;
    in dzadaob, 65, 74; in gardens/
    houselots, 5, 86, 190–91; preserva-
    tion in, 42; quality of, 5, 42, 45, 53,
    60, 62; in *rejolladas*, 64; samples, 53;
    wetland, 45, 46
Solares (houselots): ancient, 4–5, 43–54,
    86–91; modern, 7, 187–97. *See also*
    Houselots
Sotuta ceramic complex, 65, 151, 152,
    160, 164, 165, 169, 170, 182
Southern Maya lowlands: agriculture,
    43–44; compared to northern low-
    lands, 3–36, 45, 143, 152–53, 155;
    iconography at Ichmul de Morley,
    167; relations with Chac II, 141;
    relations with Teotihuacan, 140
South Pyramid at Chac II, 126, 133, 134
Spanish conquest. *See* Conquest
Stanton, Travis, 5, 22–23, 25, 27, 30, 33,
    34
Stelae, 17–18, 32, 98, 125–38, 143, 147,
    166, 171. *See also* Monuments
Stephens, John Lloyd, 42, 107
Stone walls, 49, 123, 134, 192. *See also*
    Albarradas
Structural Core, 188–91, 195–96
Suhler, Charles, 150

Surface collection, 20, 65, 67, 114, 120, 160

Surveys: botanical, 194; at Chac II, 121; at Ek Balam, 66–76; hinterland, 5, 42, 46–51, 58–60, 61, 65–69; at Ichmul de Morley, 160; of natural resources, 60, 61–64; regional, 15, 20, 26, 34, 57, 156; at Sayil, 120; topographic, 41, 52–53; wetland, 46–51. *See also* Settlements

Talud, 20, 133, 138

Tancah ceramic group, 100, 103–5, 110–14, 149. *See also* Saban, ceramic group

Taube, Karl A., 99, 110–12, 179, 201

Temple of the Warriors, Chichén Itzá, 178–79

Teotihuacan (Mexico): control of Copán, 119; control of Tikal, 119, 145; influence at Chac II, ix, 6, 119–23, 129–41; trade, 140. *See also* Teotihuacano

Teotihuacano, 104, 139, 147, 152

Terminal Classic period: ceramics, 169; at Chac II, 117, 121, 139–41; at Chichén Itzá, 177; at Chunchucmil, 78; at Ek Balam, 65, 163–65, 169; florescence in the northern lowlands, 16–17, 36; at Ichmul de Morley, 159; interaction spheres, 97; Megalithic architecture, 32, 101, 102, 107, 113, 117; political change, 146; political organization, 66; at Victoria, 113; at Yaxuná, 146, 151–53; at Yulá, 169

Tezcatlipoca, 175–76

Thin Orange ceramic, 124, 136–38

Thin Slate wares, 136, 151

Thompson, J. Eric S., 174, 176

Tikal (Guatemala): architecture, 109; ceramics, 145; control by Teotihuacan, 119, 136–39, 145; Early Classic, 109; mapping project, 56; obsidian, 139; regional control by, 142; settlement patterns, 86

Tikintzec (Yucatán) site, 70–74

Timucuy ceramics, 104, 105, 111, 113, 114

Tipikal ceramic, 26, 102, 105, 113

T'isil (Quintana Roo), 50, 51, 55

Tlaloc, 149, 176, 181

Tohil Plumbate ceramic, 108, 164, 176, 181, 182

Toltec: at Chichén Itzá, 7, 176–83; not at Ek Balam, 166–70

Toltec Chichén. *See* Chichén Itzá

Tombs, 30, 111, 147. *See also* Burials

Tortillas, 216–17, 219

Tourism: and archeologists, 199–200, 206–7; effects on modern Maya, x, 7, 8, 199–219

Trade: in ceramics, 145, 147, 149, 151, 164; at Chac II, 6, 120, 137, 139–40; at Chunchucmil, 33–34, 78–79, 91–92; Early Classic, 33–35, 147; and interaction spheres, 96; Late Classic, 149; Middle Classic, 149; Middle Formative, 23–25; of obsidian, 136; routes, 78, 169; Terminal Classic, 151; at Yaxuná, 115, 147, 149, 151

Transects: between Chichén Itzá and Ek Balam, 156–57, 164; at el Edén, 48, 52–53, 54; as survey method, 48, 59, 156; at Yohdzadz near Ed Balam, 67. *See also Brechas*

Tres Lagunas (Quintana Roo) site, 104, 106, 108, 112

Tres Zapotes (Veracruz) site, 24

Triadic groupings: as temporal markers, 97, 100, 101; and Megalithic architecture, 100, 102, 103, 104, 105, 107, 108; at Yaxuná, 27

T-tests, 87–88

Tukil (Yucatán), 70–72

Tula (Hidalgo) site, x, 7, 173–83. *See also* Chichén Itzá; Toltec

Tulum (Quintana Roo) site, 16, 22, 81

Tumben-Naranjal (Quintana Roo) site. *See* El Naranjal

Ucí (Yucatán) site, 28, 32, 105, 106, 108, 113

Ucú Black ceramic, 105, 113

Unto ceramic, 100, 105, 113

Urban centers: ancient centers, 28, 31, 56–60, 68, 78–79, 140, 156, 194; modern development of, 210, 212, 214, 217

Uxmal (Yucatán) site, 13, 77, 143, 153

Valladolid, 61, 63, 65, 74, 204, 212

Valladolid Bichrome ceramic, 27, 103

Velázquez Morlet, Adriana, 23, 115

Venenera ceramic forms, 124, 135

Veracruz, 24, 138, 187

Victoria (Quintana Roo) site, 32, 105, 106, 108, 113

Vision Serpent, 130, 132

Warfare, 15, 96, 140, 142, 145, 149, 152, 169, 177

Water: and cenotes, 50, 61, 62, 63, 69, 73, 213; channels, 44; jars, 124; levels, 5, 52, 63; mollusks, 46, 53–54; potable, 62, 73; screening, 53; sources, 6, 45, 56, 60–65, 67, 69–70, 73–76, 139; table, 62, 63, 64; in wetlands, 52. *See also Aguadas; Cenotes*; Wells; Wetlands

Wells, 61, 64, 70, 73. *See also* Water

Western Cehpech ceramic sphere, 151

Wetlands: agriculture, 44–45, 51; manipulation, 4, 45, 46; resources, 5, 50, 55; sites near, 5, 51, 156; in Yalahau region, 41, 44–46, 52. *See also Savannas*

Wilk, Richard, 84, 87, 199, 208

Xanabá Red ceramic, 102, 103, 105, 113

X-Huyub (Yucatán) site, 27, 71, 72, 73

X-Kukikan. *See* Kukulkan

Xtobó (Yucatán) site, 19, 20, 25

Xuilub (Yucatán) site, 19, 67, 68, 70, 71, 72, 73, 74, 169

Yalahau Region (Quintana Roo): architecture, 32; caves, 20; in the Late Formative period, 26; rainfall, 46; sites, 32, 49; terrain, 52; wetlands, 44–46, 51, 55

Yalahau Regional Human Ecology Project, 15, 41, 45, 58, 156, 201

Yaxhom (Yucatán) site, 32, 34, 100, 106, 108, 114

Yaxuna: IIb, 147, 154; III, 111, 149, 150, 151; IVa ceramic complex, 151, 152; IVb ceramic complex, 151

Yaxuná (Yucatán): architecture, 27, 30, 31–33, 108, 115, 150, 151; burials, 115, 147; ceramics, 20, 24, 111, 115, 147, 149, 151, 152, 165; and Chichén Itzá, 147, 151–52, 153; and Cobá, 146, 149, 150; in the Early Classic, 34, 111, 147; in the Late Classic, 149; in the Late Formative, 27, 30; Megalithic style, 31, 32, 33, 108, 115, 147; in the Middle Classic, 149; in the Middle Formative, 20, 23; Puuc style, 151; *sacbe*, 115, 146, 150; site, 6, 106, 114, 143, 154, 169; stela, 147; and Teotihuacan, 152; in the Terminal Classic, 151, 153

Yaxuná Archaeological Project, 15, 115, 146, 147, 153

Yodzil (Yucatán) site, 70, 71, 72

Yohdzadz (Yucatán) site, 67, 68, 70, 71, 72, 73

Yucatán Atlas Project, 156, 160. *See also* Garza Tarazona de González, Silvia; Kurjack, Edward